THE NATURALIST'S ILLUSTRATED GUIDE TO THE SIERRA FOOTHILLS AND CENTRAL VALLEY

THE NATURALIST'S ILLUSTRATED GUIDE TO THE SIERRA FOOTHILLS AND CENTRAL VALLEY

DEREK MADDEN

WITH KEN CHARTERS AND ERINN MADDEN
ILLUSTRATED BY DEREK MADDEN

HEYDAY
BERKELEY, CALIFORNIA

Library of Congress Cataloging-in-Publication Data

Names: Madden, Derek, author, illustrator. | Charters, Ken, author. | Madden, Erinn, author.
Title: The naturalist's illustrated guide to the Sierra Foothills and Central Valley / Derek Madden with Ken Charters and Erinn Madden ; illustrated by Derek Madden.
Description: [Revised edition]. | Berkeley, California : Heyday Books, [2020] | First Heyday edition 2005, originally titled Magpies and Mayflies: An Introduction to Plants and Animals of the Central Valley and Sierra Foothills.
Identifiers: LCCN 2019057785 (print) | LCCN 2019057786 (ebook) | ISBN 9781597144865 (paperback) | ISBN 9781597144971 (epub)
Subjects: LCSH: Natural history--California--Central Valley.
Classification: LCC QH105.C2 M15 2020 (print) | LCC QH105.C2 (ebook) | DDC 508.794/5--dc23
LC record available at https://lccn.loc.gov/2019057785
LC ebook record available at https://lccn.loc.gov/2019057786

Illustrations by Derek Madden

Cover Art: Derek Madden
Cover Design: Ashley Ingram
Interior Design/Typesetting: Ashley Ingram

Published by Heyday
P.O. Box 9145, Berkeley, California 94709
(510) 549-3564
heydaybooks.com

Printed by Lightning Source, USA
10 9 8 7 6 5 4 3 2

To Sierra and the World of Wonder

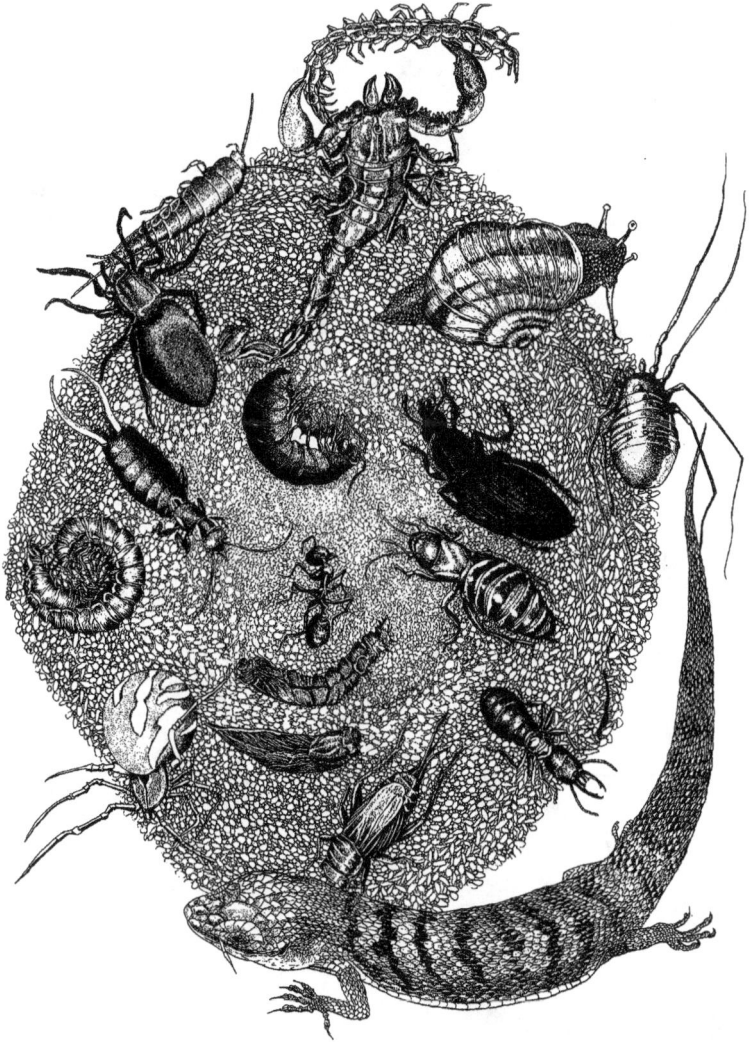

CONTENTS

PREFACE

California's central valley is a vast region, spanning nearly 450 miles, where waterways crease a landscape of grasslands, orchards, and cities, from Bakersfield in the far south to Redding in the north. Rising up from this vale is a rim of rolling foothill savannahs that resemble the great upland plains of East Africa. A break in this rim lies at the Delta, where a labyrinth of braided waterways, islands, and marshes mark the flow of inland rivers colliding with the waters of the Pacific ocean. This land of little rain was occupied in prehistoric times by enormous inland seas and later fostered great civilizations of Native Americans. In more recent times, fieldworkers and landless farmers immigrated to this vale to escape Midwestern dustbowls, a pattern repeated by newer arrivals from across our Southern border and even from across oceans. They've come to scrape together a life in the booming croplands and orchards and on the railroads and trucking routes that connect this fertile hub to its far-flung markets.

Today, the predominant landscape of the Central Valley and Sierra Foothills is still broad spreads of field and tree crops, though now the farms are often edged by housing estates. Suburban development has come to the Sierra foothills as well. Throughout the Valley/Foothill region, grazing livestock, fires, and the introduction of plants and animals—some invasive, some less so—have altered the native landscape irreparably. Yet California's central valley is a place that continues to teem with life: the Valley is a migratory corridor for birds on the Pacific Flyway and for several species of fish; habitats for bizarre creatures from horned lizards to kangaroo rats to freshwater jellyfish abound. To this day, the deep riverside forests and vast oak savannahs of the region resound with the cries of woodpeckers and the screeches of hawks.

The Naturalist's Illustrated Guide to the Sierra Foothills and Central Valley introduces the plants and wildlife of this enormous region. We include facts that delight and astound: did you know that an otter has a thousand hairs per square inch of skin, or that lizards can detach their tails in order to distract predators? The tails of some species can continue to wriggle for up to an hour. To some extent, this is a historic guidebook, describing a fabulous array of creatures that once roamed in great numbers and still exist where they can in backyards, vacant lots, and reserved parks across this land.

A note of caution: eating wild plants and fungi is inherently risky. The publisher and authors are not responsible for any adverse effects or consequences that might result from the consumption of toxic wild substances. Information contained in these pages is based on scientific literature and several decades of our own fieldwork, and it has been reviewed by experts in various fields of science. However, our approach here is to showcase stories of the marvelous life forms that inhabit this region in order to share this ecological diversity with all those who treasure their wild neighbors here in central California.

Among the many people who have helped us are David Aurora, Harold Basey, Arnold Chavez, Teri Curtis, Sarah Davis, Tana Dennen, David Grubbs, Lynn Hansen, Tim Heyne, Carl Johansson, Maxine Madden, Sierra Madden, David Martin, Malcolm Margolin, Elizabeth McInnes, Cathy Snyder, staff at the Great Valley Museum, Catherine Tripp, Lillian Vallee, Guy VanCleave, and Criss Wilhite. David Lukas, for thoughtful comments on the Heyday edition, deserves special mention, as do editor Jeannine Gendar, designer Ashley Ingram, and art director Diane Lee. Thanks to Marthine Satris for her editorial precision and creativity during publication production. And for their generous support of this project, we thank James McClatchy and the Strong Foundation for Environmental Values.

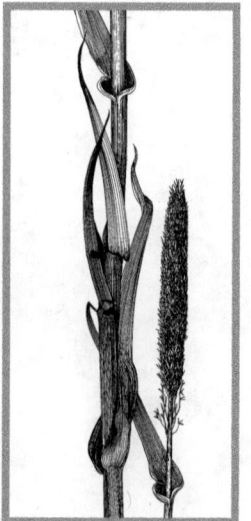

PLANTS

Plants with Spores | Plants with Naked Seeds |
Flowers, Fruits, and Seeds | Leaves, Roots, and
Stems | Surviving Fire | Surviving Deadly Salts
| Surviving in Water | Vernal Pools | Poisons,
Thorns, and Galls | Wildflowers and Weeds

Plants with Spores

As different as ferns and mosses appear to be, they share the same basic need: water, and lots of it. Anchored in dampness near leaky gutters and riverbanks, they thrive on the shady side of life. Their ancestors may have fed dinosaurs, but those larger plants disappeared along with the prehistoric swamps; in our dry habitats today, ferns and horsetails rarely grow over four feet tall. Mosses don't grow much taller than a shag carpet.

The reason ferns are so much larger than mosses lies in the vascular system that transports water and nutrients within a plant. Although seedless and primitive, ferns have a network of vascular tubes that enable large growth. Rootless and lacking a good vascular system, mosses get a limited amount of water because they must absorb it through whatever part of the plant is wet.

When moss plants are wet, their specialized leaf cells swell like water balloons; moss holds nearly twenty times its dry weight in water, which puts cotton to shame when it comes to absorbency. Because of their legendary water-holding capacity, mosses have been used as bandages and for gardening soil in many cultures. Sterilized moss was the nurse's bandage of choice during World War I. It was gradually replaced as hospitals switched to white cotton for its reassuringly clean appearance.

Ferns, horsetails, and mosses—seedless green plants—rely on an odd means of reproduction involving spores. Once released from the plant's sporangium capsule, spores grow into multicellular masses called gametophytes. These produce either sperm or eggs that join to produce a young plant that grows into an adult.

Native Americans observed that animals do not eat most types of ferns, an indication of toxicity, and alkaloids and inorganic toxins are indeed found in many species. The ribbed stems of horsetails contain gritty silicon dioxide, rendering these plants all but inedible, but that doesn't mean they are useless. Miners named this plant the "scouring rush" during the gold rush because it was handy for scrubbing cooking pots.

Sporangium and spores on leaf

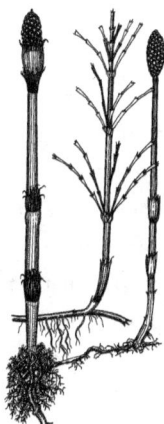

Horsetails
(*Equisetum* spp.)

Adult fern

Young fern growing from the
gametophyte stage

Classification note: Ferns, horsetails, and seed-producing plants are in division Tracheophyta; mosses are in division Bryophyta.

Plants with Naked Seeds

Pines and their relatives have naked seeds; because they lack the protective rinds and shells of walnuts, cashews, and many other fruits, these plants armor their vulnerable seeds with cones.

Wind is the tool that most of these species use to distribute pollen to the large female cones. A yellow dusting of pollen is a common sight in fall, sometimes even when the nearest pine is far from sight. This shotgun approach to fertilization has a dubious success rate, so in most species, each tiny male pine cone produces hundreds of pollen grains. Bubble-like air bags or wings keep the microscopic pollen aloft over long distances. Upon landing on a female cone, the pollen discharges sperm and a fertile pine nut develops.

Resin and pitch are among the notable features of pines: gaseous terpenoids in resin evaporate when exposed to air, causing the resin to become a gooey pitch. Where the tree is injured, resin flows outward to form a scab before infection sets in.

Pine nuts are loaded with carbohydrates, fatty lipids, and some amino acids. With so many nutrients packed into each cone, pine trees help sustain the lives of various animals and were a staple of the Native American diet in some areas. People gathered unripe cones, probably to beat squirrels and other animals competing for the fatty nuts. Roasting cones over a fire helps loosen the nuts by melting resins that glue the green pine scales together.

Classification note: Pines and other naked seed plants are in class Pinopsida.

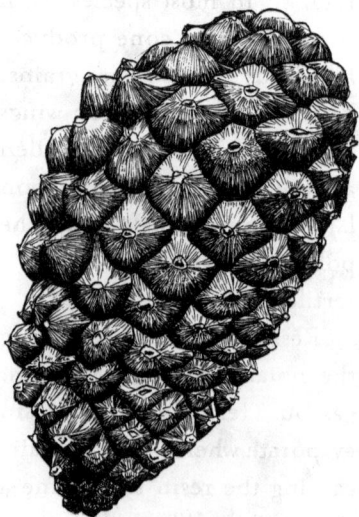

Knobcone pine
(*Pinus attenuata*)

Flowers, Fruits, and Seeds

Plants cannot uproot themselves to search for a mate when it is time to reproduce. Instead, the anthers of a flower pack sperm into pollen grains that can survive a trip to another flower's ovary. Pollinators—bees, butterflies, hummingbirds, and other creatures that transport pollen—are nature's matchmakers. And the life cycle and structure of a flower fit its target pollinator like a catcher's mitt fits a baseball. The colors of flowers, their markings, their odors, and their nectars are all designed to entice pollinators.

In the eyes of bees, the world is a bleak land of grays, except for blue, violet, and ultraviolet, which they see vividly. Flowers with these color markings are targeted by such bees. Several species of butterflies are similarly attracted to oranges, while moths, are drawn to white flowers during their nocturnal forays. Spicy-smelling flowers attract several species of leaf beetles. Flowers that stink often entice pollinating flies to visit. Hummingbirds are attracted by red flowers.

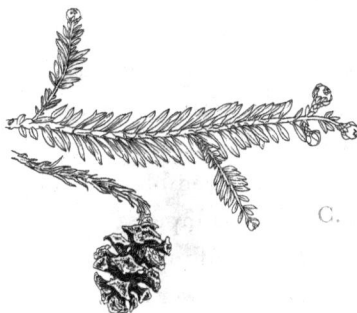

PINES AND THEIR RELATIVES

A. Incense cedar (*Calocedrus decurrens*): The fragrant reddish wood is used in making pencils and furniture.

B. Big tree (giant sequoia, *Sequoiadendron giganteum*): California's heaviest trees arise from seeds in cones the size of chicken eggs.

C. Redwood (*Sequoia sempervirens*): Tiny seeds from olive-sized cones sprout into California's tallest trees. The redwood doesn't occur in the Valley/Foothill region unless planted.

D. Ponderosa pine (*Pinus ponderosa*): The pineapple-sized cones are prickly.

E. California foothill pine (*Pinus sabiniana*): Appears grayish-green from a distance. Huge spiky cones.

Small, crusty, and otherwise boring flowers may not attract animal pollinators, but the job of fertilization still takes place. Such plants often perch their drab flowers so that wind or water will transport pollen to prospective mates. Wind is a cheap pollinator, requiring little of the plant's energy, but wind-pollinated plants have to produce incredible amounts of pollen to compensate for the wind's lack of precision.

Once a flower has been fertilized and then made the dramatic change to being a fruit, it must protect itself from being eaten by the very animal it will later need to disperse its seeds. Some fruits—the juicy ones—manage this is by storing complex carbohydrates and bitter-tasting natural metabolites until the seeds within are mature. A seed's embryo feeds on its own food storage, so the surrounding fruit can become toxic or tough without harming the seed. A camouflage of green or brown also protects young fruits from being noticed and eaten.

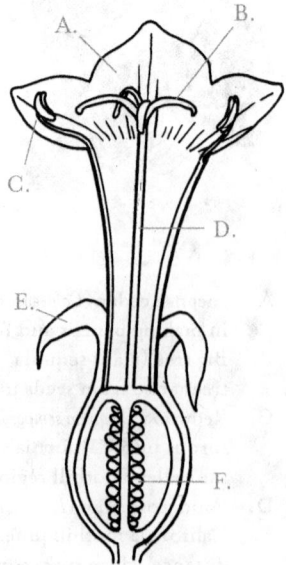

FLOWERING PLANT
(SUBDIVISION
SPERMATOPHYTINA)

A. Petal (plural is corolla)
B. Stigma
C. Anther (top of stamen)
D. Style
E. Sepal (plural is calyx)
F. Ovary (ovules inside)

SHRUBS AND VINES WITH JUICY FRUITS

A. Blackberry (*Rubus* spp.): Himalayan blackberry, a non-native, grows like a weed into prickly mounds. California blackberry has smaller prickles and fruits. (Rosaceae)

B. Gooseberry and currant (*Ribes* spp.): Trumpetlike flowers arise from a swollen ovary. (Grossulariaceae)

C. Wild rose (*Rosa californica*): Prickly plants with pinkish flowers and orange fruits. Leaflets in groups of five or seven. (Rosaceae)

D. Blue elderberry (*Sambucus nigra*): White to cream-colored flowers become black berries in clusters. (Caprifoliaceae)

E. Thimbleberry (*Rubus parviflorus*): White to pinkish flowers become raspberry-like fruits. (Rosaceae)

F. California wild grape (*Vitis californica*): Wild grapevines with peeling bark and tendrils. (Vitaceae)

As seeds mature, the surrounding fruit converts inner compounds into tasty carbohydrates. Sweet odors and attractive outer colors alert animals that a fruit is ripe enough to eat. Seeds are ready to germinate after having passed through an animal's digestive tract. The seeds of uneaten fruit are often killed by insects before they get a chance to sprout.

The California black walnut (*Juglans californica*) was often planted by the Central Valley's Yokuts people.

Oaks and their hard fruits are one of nature's success stories. A hard, woody inner ovary wall protects the oak embryo until it is ready to sprout. Bitter-tasting tannin in the outer wall further discourages seed predators from nibbling on the fragile embryo.

Well over fifty thousand creatures inhabit an oak during its lifetime. Acorn death begins when the nuts are still on the tree and are under attack by larval filbert worms and weevils. Acorn woodpeckers chisel holes in tree trunks and then fill them with acorns. Woodpeckers can harvest 8 percent of the year's acorn crop, and rarely does a bird drop a nut that survives to become a tree. Scrub jays gather and store many acorns, but they forget or don't need 50 to 70 percent of the nuts they hide, so they accidentally plant many oak trees. Rodents, deer, and livestock also eat acorns. Sprouted acorns are often killed by deadly root fungi that spread when land around an oak is watered in summer.

Acorns contain a nutritious blend of carbohydrates, lipids, amino acids, phosphorus, and calcium. Add to this the staggering number of acorns that fall each year, and the immense influence that oaks have on the ecology of an area becomes clear. A mature valley oak dumps nearly five hundred

The blue oak's (*Quercus douglasii*) leaves are bluish-green with wavy edges.

pounds of acorns to the ground in a good year. Even the skimpy **canyon live oak** (*Quercus chrysolepis*) yields about two hundred pounds per year.

Despite appearing to be rotting where they stand when they reach old age, oaks are tough trees that take a long time to die of old age. If they escape a premature death, **valley oaks** (*Quercus lobata*) may live for six hundred years. Even when a venerable tree dies, it continues to provide habitat until it finally rots away.

Acorns were the nutritional cornerstone of native California. Native people husked acorns and used pestles to crush the inner nuts in rock mortars. They then put acorn meal into baskets or holes lined with sand and poured water through the meal to leach out tannins. They often made soup or mush with the acorn meal by boiling it in baskets with stones heated in the fire. The watertight baskets used in processing acorn mush are marvels of functional art and design.

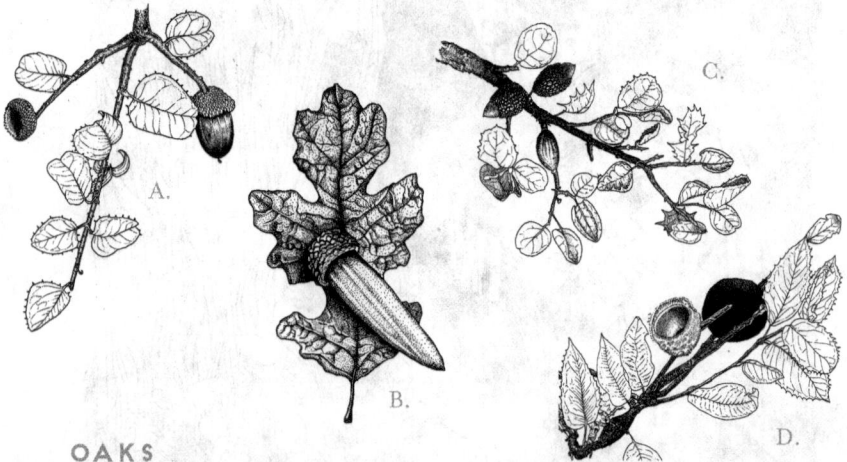

OAKS
(FAMILY FAGACEAE)

A.   Scrub oak (*Quercus berberidifolia*): Shrub with many base stems and prickly leaves.
B.   Valley oak (*Quercus lobata*): Leaves have several lobes with rounded tips. Acorns are long.
C.   Interior live oak (*Quercus wislizeni*): This oak has a main stem with a dense growth of prickly evergreen leaves.
D.   Canyon live oak, or gold cup oak (*Quercus chrysolepis*): Acorn cup and underside of leaf are fuzzy yellow.

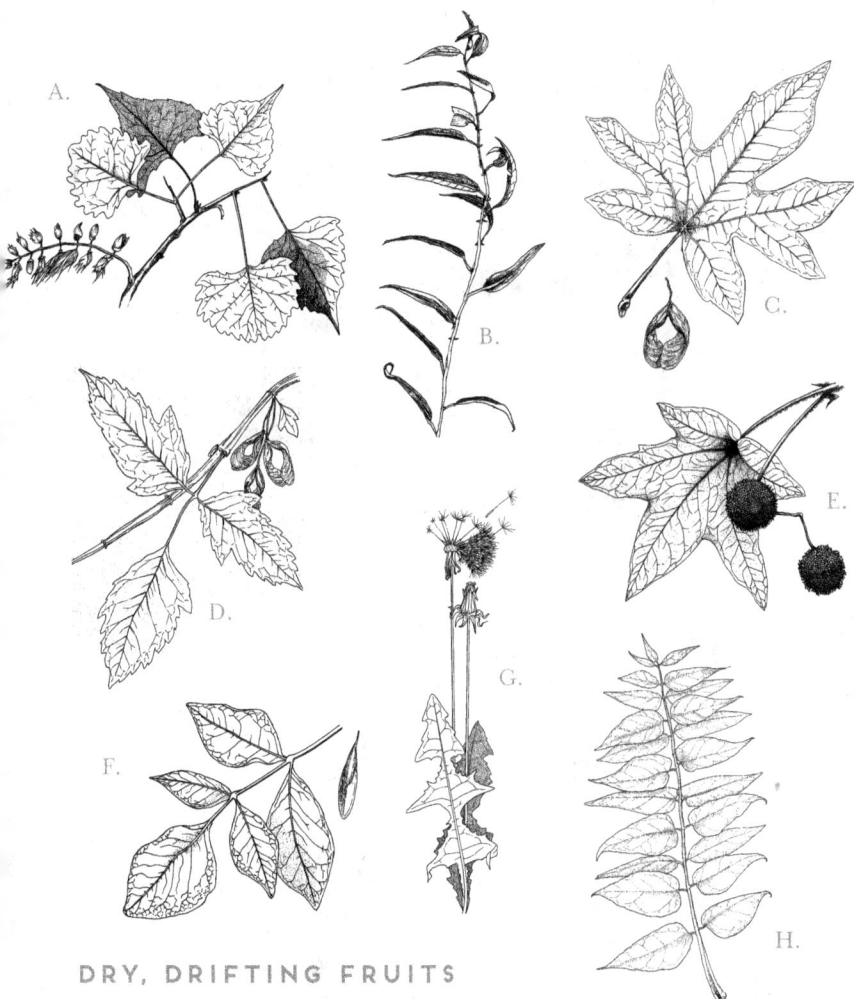

DRY, DRIFTING FRUITS

A. Fremont cottonwood (*Populus fremontii*) and
B. Willow (*Salix* spp.) produce clumps of cottony tufted seeds in fall. (Salicaceae)
C. Big leaf maple (*Acer macrophyllum*) and
D. Box elder (*Acer negundo*) produce dry fruits that twirl when falling. Maple family (Aceraceae)
E. Western sycamore (*Platanus racemosa*) produces a rounded seed head filled with tiny lightweight seeds. Sycamore family (Platanaceae)
F. Oregon ash (*Fraxinus latifolia*) has clusters of dry winged seeds that twirl when drifting in the wind. (Oleaceae)
G. Dandelion (*Taraxacum officinale*) is a common lawn weed with yellow ray petals. (Asteraceae)
H. Tree of heaven (*Ailanthus altissima*): Railroad workers planted this tree from China for shade in California's early days of settlement. With the help of prodigious amounts of drifting fruit, it has since spread like a weed. (Simaroubaceae)

The Miwok people found cotton-wood's soft wood ideal for carving gadgets such as the pump drill. When this device is tipped with sharp obsidian, its smooth action allows it to drill through hardwoods and even soapstone easily.

The massive cottonwood forests of the past disappeared when rivers were harnessed by dams; the sandy beaches where cottonwoods once germinated are now overgrown with willows and weedy vegetation. Some gnarled old cottonwoods survive today, standing like prehistoric statues in California's interior.

Soaproot (*Chlorogalum pomeridianum*)

Underground, beneath the wavy leaves of the soaproot plant, is a modified stem that looks like a big bar of soap surrounded by stiff brown fibers. Native Americans used digging sticks, and the proper balance of muscle and skill, to coax these meaty soaproot corms from deep within the earth.

This member of the lily family (Liliaceae) stores its energy as carbohydrates and lipids underground, where few animals can reach this potential food supply. It is a Miwok tradition to stuff game balls and children's dolls with the fibers that surround soaproot corms, and these fibers are also bound together into brushes used for a variety of purposes. Because soaproot corms contain saponin chemicals, which dissolve many compounds, the waxy core of soaproot, once stripped of its fibrous envelope, can be used as a mild but effective hand soap.

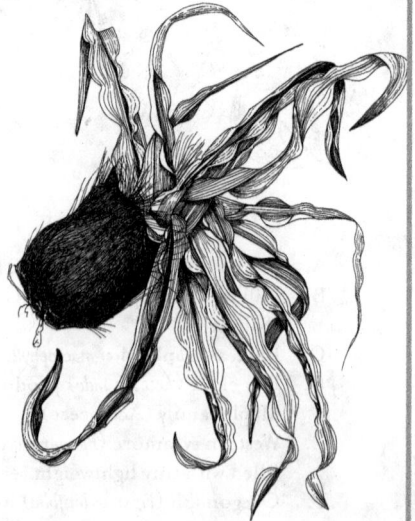

The survival strategy of drifting fruits differs from that of juicy fruits. Winged fruits that twirl as they fall seem graceful if pathetic, but they don't need to drift far, just far enough away from the competitive roots of their parents. Cottony fruits work in a similar manner, but they cover more ground. Cottonwoods and willows growing along rivers make use of the frequent upstream breezes to carry their fruits to promising shores. Sometimes a wayward seed floats overland to colonize new habitats. Regardless of the method used, most drifting fruits mature in late summer, just in time for the breezy days of early fall. Cottony tufts and twirling fruits are among nature's signs that another summer has ended.

Leaves, Roots, and Stems

A simple thing can sometimes perform miracles. This is true of the green chlorophyll pigments in leaves. The miracle begins when sunlight hits water stored in a leaf. Oxygen shoots off from the reaction to become a vital part of the air we breathe. Meanwhile, the carotenoid pigments that surround the chlorophyll absorb sunlight. This energy is transferred to chlorophyll, which bonds hydrogen, oxygen, and carbon into an organic nutrient, glucose. In the fall, chlorophyll dies, and the brilliant orange and yellow carotenoids are unmasked for all to see.

Leaves have many tiny stomata in their skin that allow gases in and out. This helps keeps a leaf alive, but it can also be lethal when too much water vapor escapes a leaf. Desert plants have small leaves and a waxy cuticle, both of which reduce water loss. Some plants even have stomata sunken deep within the skin of each leaf. Very little water vapor diffuses out of these damp stomata caves, and the plant can hold on to its life-giving water until the rains arrive.

Roots and stems work together to bring water and minerals to a plant's leaves. Most of the wood of a tree's trunk is composed of xylem tissue, which delivers liquids upwards through vessels and straw-like cells (tracheids). After leaves have bonded glucose sugars together, the stem must go to work again, this time in the opposite direction. Outside of a tree's core of xylem, a tissue of tubes called phloem functions as the tree's bloodstream. Sugars are delivered throughout the plant, even into the roots, to feed every living cell. Roots and stems also store substances that the plant may need later.

In spring, when water and nutrients are abundant, the xylem in a tree grows broad. Later in the year, the tree lays down a layer of xylem called latewood. This dark, compressed xylem made in summer tells a tale of summer drought and harsh conditions. Year after year, these rings of growth pile up one after another, which is why the age of a tree can be told by counting them.

Surviving Fire

Sweeping through the landscape at temperatures approaching 2,000°F, wildland fires convert life into ash and smoke within minutes. Yet there are plants in California that manage to thrive under such conditions. Many of these pyrophilic (fire-loving) plants are unrelated, but all share the quality of having adapted to the natural rhythm of fire.

First to appear in a scorched landscape are herbs that blow in as seeds and germinate in the nutritious ash. Their reign is brief, especially if there are crown-sprouting plants around. Among the champions of crown sprouters are **chamise** (*Adenostoma fasciculatum*) and **scrub oak** (*Quercus berberidifolia*), both of which survive fires as underground roots. These shrubs don't have to wait for the first rains after a fire to begin growing, for their roots have their own water supply. Other shrubs, such as **manzanita** (*Arctostaphylos* spp.) and **ceanothus** (*Ceanothus* spp.), survive fires by producing rock-hard fruits that can lie in the ground for several years, unable to sprout until scorched by a fire.

Stalwart and amazingly tough are a group of pyrophilic trees that survive fires by taking them head on. Sequoia does this with the help of its extremely thick bark, which

FIRE-LOVING PLANTS

A. Manzanita (*Arctostaphylos* spp.): Manzanita is recognizable by its fruits, which look like tiny apples, and its red bark that often peels off. (Ericaceae)

B. Coyote brush (*Baccharis pilularis*): The evergreen leaves of coyote brush shine with resins. (Asteraceae)

C. Mountain mahogany (*Cercocarpus montanus* var. *glaber*): The coiling styles are long and hairy. (Rosaceae)

D. Scrub oak (*Quercus berberidifolia*): This prickly shrub produces small acorns. (Fagaceae)

E. Yerba santa (*Eriodictyon californicum*): Leaf resins cause yerba santa to be sticky and smell something like turpentine. The flowers are lavender or white. (Hydrophyllaceae)

F. Chamise (*Adenostoma fasciculatum*): The long branches of chamise are covered with narrow green leaves that look like small pine needles. (Rosaceae)

may blacken but burns slowly. One of the reasons for this is that sequoia bark has no flammable resins. Another is that the fibrous weave of sequoia bark smolders slowly rather than producing the large flames typical of, for example, a burning pine tree.

Surviving Deadly Salts

The mineral deposits that evaporating water leaves behind on an old drinking glass are harmless, but such deposits in soil can be lethal to plants, especially where puddles evaporate in the same dry landscapes year after year. Mineral salt loads can turn fertile soils into harsh alkali flats over time. Salts draw water outward from the roots—the wrong direction for a thirsty plant. Once inside the plant, mineral salts kill by disrupting water balance in much the same manner. But to a halophyte, a plant adapted to life in a saline environment, such mouth-puckering conditions are no big deal.

Pickleweed (*Salicornia* spp.) and **iodine bush** (*Allenrolfea occi-*

dentalis) survive in alkaline soil by concentrating salty water in the tips of their branches. This burden of salt is dumped each time a dead tip drops off the plant. They always seem to be shedding branch tips, and for good reason. **Saltbush** (*Atriplex* spp.) has a simpler approach: salt oozes through channels of stomata in the skin of its leaves and then rain or wind washes it off. You can sometimes see a white crust of salt right on the leaf of a saltbush.

Halophytes have clever survival tricks, but they still struggle to stay alive. Many grow widely spaced in the landscape so their roots can harvest precious rainwater without having to share it with a neighboring plant.

Alkali flats are often shown in movies as a dusty land of gaunt cattle, worn-out cowboys, and tumbling tumbleweeds. In reality, cattle and the prickly weeds are recent arrivals, and these habitats are alive with kangaroo rats, leopard lizards, and a lot of other bizarre native wildlife. Tumbleweeds made their first California appearance in the early 1900s, after seeds contaminated

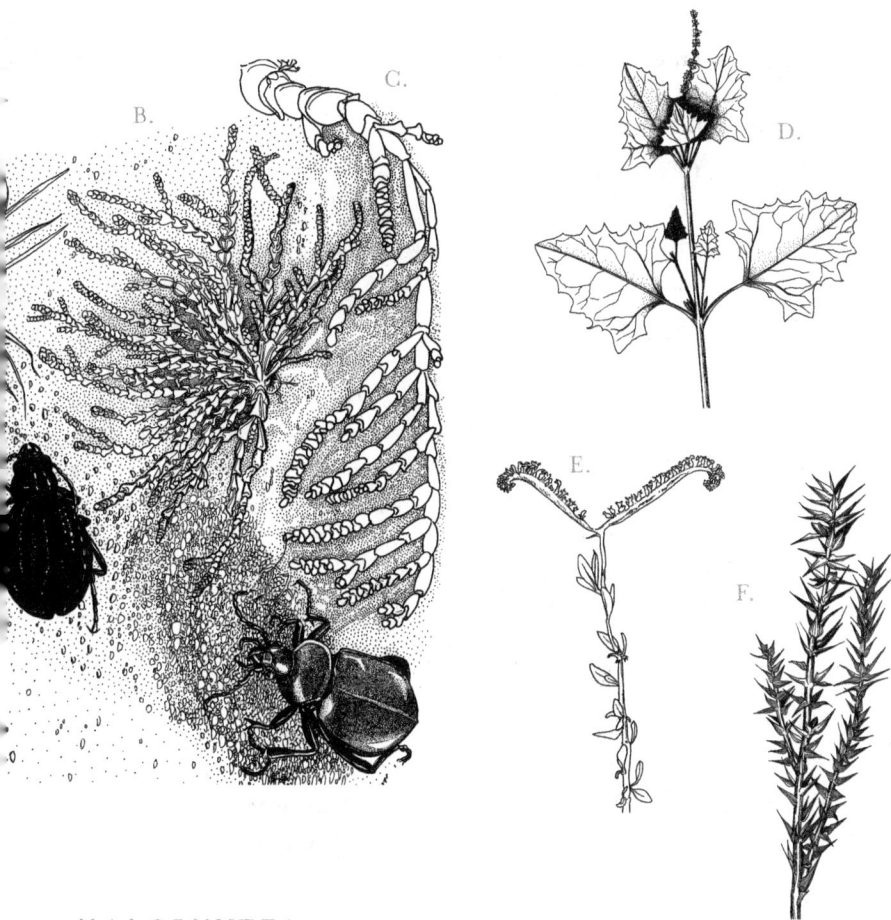

HALOPHYTES

A. Pricklegrass (*Crypsis* spp.): This grass is often prostrate or creeping. For a grass, the leaves are fairly short. The leaf sheath and ligule are hairy. (Poaceae)

B. Iodine bush (*Allenrolfea occidentalis*) and

C. Pickleweed (*Salicornia* spp.): These two plants are somewhat similar, both having young stems and leaves that are juicy like pickles. Stem tips are purplish. The flowers have no petals. (Chenopodiaceae)

D. Saltbush (*Atriplex* spp.): There are many different species of saltbush, but all lack petals and have 1 to 5 sepals. Leaves may have a crust of salt. (Amaranthacaceae)

E. Salt heliotrope (*Heliotropium curassavicum*): The tiny flowers of this herb blossom along a coiling stalk. (Boraginaceae)

F. Tumbleweed, or Russian thistle (*Salsola tragus*): Thin, green stems harden and become prickly. This weedy plant spreads seeds as it rolls. (Amaranthacaceae)

a grain shipment from Europe to the East Coast. The weedy shrubs rolled their way across the states to somehow find a place in tales of gunslingers and the Wild West.

Surviving in Water

From shoreline inhabitants to those that spend their entire lives underwater, different plants tolerate water differently. Getting enough oxygen is one of the biggest challenges for plants in damp habitats, because waterlogged soils contain low levels of this vital gas. Heavy bacterial action in water can drop free oxygen down to dangerous levels for a submerged plant.

Life in water is not for the random seed that happens to fall there; each species of aquatic plant has nifty devices that make the aquatic life possible. For example, the stems of cattails and some sedges contain air cavities that help channel oxygen to roots buried in soggy soil.

Anchored in mud and swaying with the current are an odd group of submarine plants called the submergents. They spread themselves mostly through stem fragments. A small piece of drifting **waterweed** (*Elodea* spp. and *Hydrilla verticillata*) may send out roots that grow several feet long before gripping a soggy river bottom.

Once rooted in place, a submergent plant erupts into a clump of individuals while also sending out runners like those of a strawberry plant. Sex, when this rare event does occur, is generally at the surface. Tiny flowers on thin stalks rise to the surface like air bubbles. There, a puff of wind delivers pollen from some flowers to others. The resulting embryos may drift away to root in a fertile, muddy bottom. Some submergents, such as waterweeds, can spread and flourish into submarine forests that trap drifting mineral sediments, gradually building soil.

"Emergent" plants are only partly submerged: anchored in mud by a meaty stem, or rhizome, the stalks of emergent plants poke out of the water. When the stalks turn brown and die each fall, the plants continue to feed by devouring

SHORELINE PLANTS
AND SUBMERGED PLANTS

A. White alder (*Alnus rhombifolia*): The red roots of this slender shoreline tree often dangle in rivers. (Betulaceae)

B. Smartweed and

C. Button bush (*Cephalanthus occidentalis*): The young stems of this shoreline shrub are red. Nutlets are clustered in round seed heads. (Rubiaceae)

D. Knotweed (*Polygonum* spp.): A leafy sheath and swollen stem form at the junction of leaves and stem. The juices of some of these plants can irritate human eyes. (Polygonaceae)

E. Horned pondweed (*Zannichellia palustris*): Opposite, threadlike leaves grow along the stem underwater. Flowers have only one stamen. (Zannichelliaceae)

F. Waterweeds (*Elodea* spp. and *Hydrilla verticillata*): Leaves are clustered in whorls. The tiny flowers float from long tubes. These submerged plants spread like weeds. (Hydrocharitaceae)

their own rhizomes. Muskrats shred rhizomes as they feed, releasing clouds of starches underwater that are eaten by plankton. Many aquatic food chains are linked to the hidden but mighty rhizome.

Changing water levels are a challenge to emergents. Spring floods would rip these plants loose if not for the rhizome's woody grip. Leaves, too, are adapted to deal with rushing water. The ribbonlike lower leaves of **arrowhead** (*Sagittaria* spp.) flap easily, offering no resistance to the wind. Broad leaves trap sunlight in the dim shadows of a riverbank.

The rhizomes and tubers of some emergents have been tremendously significant in human diets. Early explorers, including Lewis and Clark, compared the flavor of baked arrowhead, or wapato, rhizomes to that of the Irish potato. **Nutsedge** tubers (*Cyperus* spp.) were roasted like chestnuts. Native people used handheld rock or wooden mortars to grind rhizomes for mush and bread. Stalks were used as well. Yokuts people gathered young **cattail** (*Typha* spp.) stems in spring and steamed them or ate them raw. In the Tulare Lake region, they made **bulrushes** (*Scirpus* spp.) into boats that were seventy feet long, with earth packed at one end to hold a cooking fire. Native people also made comfortable sleeping mats and clothing from the supple young stalks of rushes, and the pliable but durable leaves of emergent plants were woven into ropes and marvelous baskets by various California tribes. Because natural deposits of silicon dioxide in the leaves resist nearly all forms of decay, some of these baskets last for decades.

Plants float with the help of oils or trapped air spaces while their roots absorb minerals suspended in water. Plankton and young fish flourish in the protective nursery created by floating plants. But, in the never-ending saga of the food chain, a heron or one of many other types of predators may well visit such a habitat for its next meal.

From a distance, **water hyacinth** (*Eichhornia crassipes*) looks like a wholesome field of cabbage, but this weedy plant from the American tropics clogs up California waterways

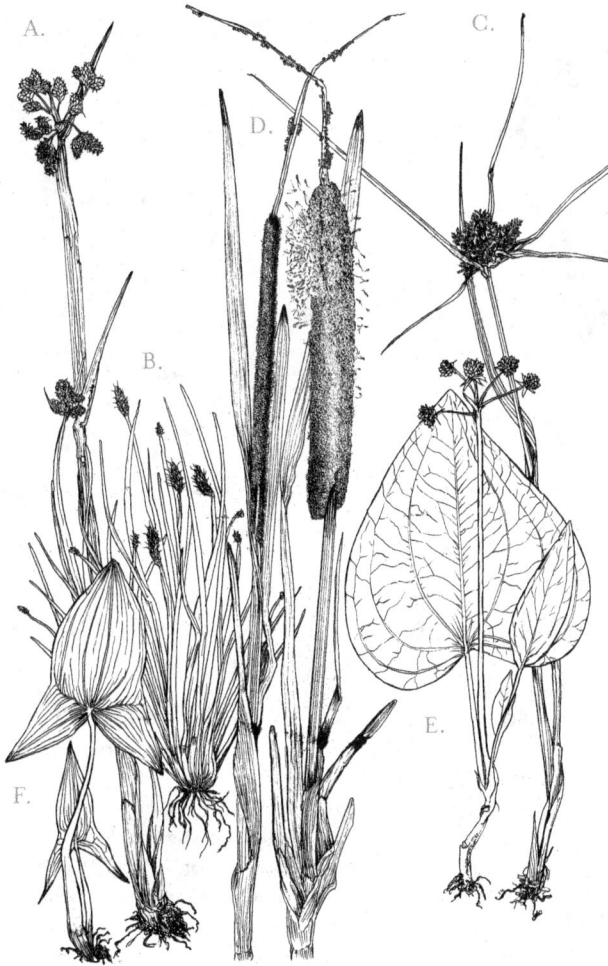

EMERGENT PLANTS

A. Bulrush (*Scirpus* spp.),

B. Spikerush (*Eleocharis* spp.), and

C. Nutsedge (*Cyperus* spp.): Most species resemble big grasses. The solid stems are angular or rounded. (Cyperaceae)

D. Cattails (*Typha* spp.): Brownish flower spikes form on erect, solid stems. (Typhaceae)

E. Burhead (*Echinodorus berteroi*) and

F. Arrowhead, or wapato (*Sagittaria* spp.): Flowers have three sepals, three white petals, and six or more stamens. Submerged leaves may be narrow, exposed leaves are broad. (Alismataceae)

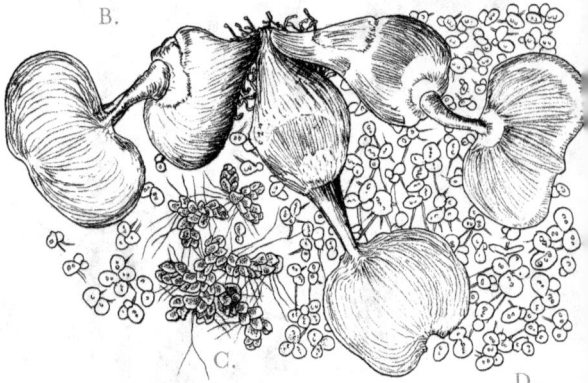

AQUATIC PLANTS

A. Water primrose (*Ludwigia* spp.): A long ovary forms beneath the four to seven yellowish flower petals. Roots may grow from stems. (Onagraceae)

B. Water hyacinth (*Eichhornia crassipes*): Clumps of broad leaves have floating, inflated stems. The large flowers of this weed range from blue to white. (Pontederiaceae)

C. Mosquito fern (*Azolla* spp.): This delicate floating fern lacks flowers, but instead makes spores in a saclike sporangium. (Azollaceae)

D. Duckweed (*Lemna* spp.) and Duckmeat (*Spirodela* spp.): These tiny plants often coat the surface of slow-moving water. Floating clusters of two or more leaves (less than 0.2 inches long) are often purplish underneath. (Lemnaceae)

E. Water lily (*Nymphaea odorata* and *N. mexicana*): The floating leaf is often hand-sized or bigger, and the lower stem is tinted red. (Nymphaeaceae)

F. Eel grass (*Zostera* spp.): This slender plant grows in the shallows of brackish estuaries near the ocean. The foot-long leaves have three to seven parallel veins. (Zosteraceae)

G. Pondweeds (*Potamogeton* spp): Pondweeds vary from broad-leaved to narrow-leaved species. The greenish flower parts are in fours, crowded into heads or spikes, often on leafless stalk. (Potamogetonaceae)

H. Ditch grass (*Ruppia* spp.)

I. Cord grass (*Spartina* spp.): This grass inhabits estuary shorelines near the ocean. (Poaceae)

J. Water nymph (*Naja* spp.)

E.

F.

G.

H.

I.

J.

A. Purple owl's clover (*Castilleja exserta*): Soft purple flowers with lower-lip petals are clustered together. (Orobanchaceae)

B. Blennosperma (*Blennosperma* spp.): There are six to fifteen yellow ray flowers on the flat receptacle of this member of the sunflower family. (Asteraceae)

C. Orcutt grass (*Orcuttia* spp.): Long leaves float when they are young, but they become rolled with age. The lemma has five teeth. (Poaceae)

D. Saltbush (*Atriplex* spp.)

E. Downingia (*Downingia* spp.): The three lower petals are fused together, as are the five stamens. (Campanulaceae)

F. Alkali mallow (*Malvella leprosa*): The leaves are rounded, and the flower buds are buttonlike. (Malvaceae)

G. Tricolor monkeyflower (*Mimulus tricolor*): The lower-lip petals are landing pads for bees. There are bright spots on the petals of this monkeyflower, one of many in the snapdragon family. (Phrymaceae)

H. Water starwort (*Callitriche* spp.): White flowers of this floating plant have two inflated bracts. (Plantaginaceae)

and smells bad when it washes ashore to die. Yet something good can also be said for hyacinth: it absorbs loads of pollutants from water, acting as a massive filter.

Mosquito fern (*Azolla* spp.) and **duckweed** (*Lemna* spp.) form a beneficial green manure that often coats the surface of water. These tiny floaters play a major role in California rice fields by generating nitrates, which all plants require to survive.

Vernal Pools

Some recurrent puddles foster a rush of radiant and bizarre life every spring and become dry basins by the end of the season. Bright rings of plants carpet the earth around these vernal pools, and the air buzzes with small bees, some smaller than houseflies, that cannot survive without the dwarf flower garden.

Each flower species in the vernal pool is visited by a specific pollinator. **Downingia** flower petals (*Downingia* spp.) are modified to act as landing pads for burrowing bees. After pollinators have made their visit, the colorful perimeter of a pool turns to brown thatch

A.

B.

C.

D.

E.

F.

G.

H.

laden with seeds that will sprout when the rains return in the next year. Seeds may remain attached to stems long after the plants die. Rains loosen the seeds so that they fall into the soil to germinate, and the vernal pool gears up for another spring.

Poisons, Thorns, and Galls

There are herbivores that would eat each plant they encountered to the ground if they could, but plants have ways to protect themselves. Toxins are one approach, sometimes with variable effects. For example, **locoweed** (*Astragalus* spp.) absorbs toxic amounts of molybdenum, arsenic, and barium from soil, so plants growing in some soils are less toxic than those growing elsewhere. Dosage also determines the effect of a toxin. Most creatures can tolerate a low dose, but a greater portion of the same toxin becomes deadly.

Because the manufacture of complex toxins leaves plants with less energy for their own growth and reproduction, the plants that develop severe toxins are those that need to protect themselves from specific, significantly harmful enemies. Cardiac glycosides and alkaloids, for example, are toxins with immediate and intense effects. Cardiac glycosides in **milkweed** (*Asclepias* spp.) are absorbed by monarch caterpillars. A bird that makes the mistake of eating a monarch experiences an increased heart rate and vomits. Such lessons teach birds to avoid these toxic butterflies. Few animals can stomach the potent alkaloids in **datura** (*Datura* spp.). **Poison hemlock** (*Conium maculatum*), another severely toxic plant, resembles wild anise and is sometimes eaten by mistake, as are the beanlike pods of toxic locoweed.

Mild toxins protect a plant to some degree, often by affecting a wider range of enemies than severe poisons. Terpenoids—essential oils and herbal extracts such as those found in mints—fit this category well. The strong odor and flavor of terpenoids reduce the appetites of everything from grasshoppers to cattle, but they are not potent enough to stop a

POISONOUS PLANTS

A. Milkweed (*Asclepias* spp.): A ring of tissue forms around the base of the flower petals. The dry fruits produce seeds with silky tassels. The cardiac toxins are fairly mild. (Asclepiadaceae)

B. Locoweed (*Astragalus* spp.): Pea flowers become inflated seed pods on this small herb. (Fabaceae)

C. Datura (*Datura* spp.): A blend of alkaloids causes datura to be toxic. Some people foolishly, and fatally, use this plant as a hallucinogen. (Solanaceae)

D. California buckeye (*Aesculus californica*): The brown fruits of this native tree contain mild saponin chemicals. (Sapindaceae)

E. Poison hemlock (*Conium maculatum*): Tiny white flowers occur in flat clusters. The greenish stems have purple splotches. This is a very toxic plant in the carrot family (Apiaceae).

F. Larkspur (*Delphinium* spp.): The dolphin-shaped flowers of larkspur are white to bluish. Delphine alkaloids make larkspur very toxic. (Ranunculaceae)

G. Henbit (*Lamium amplexicaule*): Tiny purplish flowers occur at regular levels along the stem. This is an example of a plant with mild terpenoid toxins. (Lamiaceae)

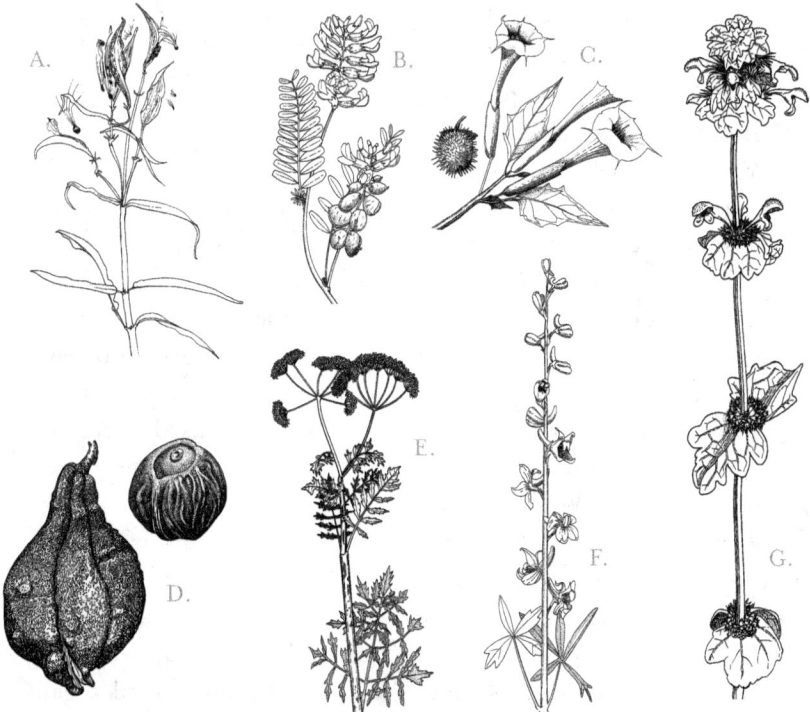

A.

B.

C.

E.

D.

F.

G.

starving herbivore in its tracks. Tannins work in an internal way to discourage herbivores by binding to digestive enzymes in the gut, thereby inhibiting absorption of nutrients and producing unpleasant results, such as diarrhea. Oaks and many other plants make use of protective tannins.

Besides toxins, plants use a remarkable array of tools to discourage herbivores. Spines that jab into flesh and thorns that hook skin discourage animals from getting too close. By comparison, bristles and woolly hairs seem like a pathetic form of plant defense, but being hairy is a popular trend in the plant kingdom. Hairs are obstacles that prevent insects from reaching the tender surface of a plant to feed or lay eggs. Eggs laid on plant hairs are doomed when the hairs are shed. Large animals experience hairs as unpleasant textures. Plant hairs can also ball up in the throat: wads of **woolly mullein** (*Verbascum thapsus*) hair have choked cattle to death. Hairs can break off a plant and cause irritation when entering lungs and eyes.

Stinging nettle (*Urtica*

spp.) injects a chemical that resembles the formic acid that ants inject when they bite. The result is all too familiar: intense pain followed by hours of itchy, irritated skin. Recovery is complete, but the lesson is learned. Nettle has trained another herbivore to leave it alone.

Few plants evoke as much fear as **poison oak** (*Toxicodendron diversilobum*) does. Swelling, redness, and blisters that form in the skin are the body's delayed allergic reaction to the urushiols of this and other plants in this genus, which affect a mammal's immune system over a period of days. Each exposure to poison oak creates a memory in the skin's mast cells, and the immune response becomes fiercer with each contact. Scratching and heat irritate the skin, causing even greater histamine release, and the tortured skin worsens. Medications that contain antihistamines can temporarily suppress mast cell sensitivity and may break the inflammation cycle.

There are plants that respond to attack by producing a knot of tissue called a gall.

SPINES, BRISTLES, AND SURFACE TOXINS

A. Milk thistle (*Silybum marianum*) and
B. Yellow star thistle (*Centaurea solstitialis*): These introduced weedy plants all have spiny flower heads. (Asteraceae)
C. Nettle (*Urtica* spp.): Tiny green flowers arise at regular intervals where opposite leaves join the stem. Nettle does well in damp, shady habitats. (Urticaceae)
D. Turkey mullein (*Croton setiger*): Another fuzzy herb, turkey mullein grows in dry and disturbed places. Doves eat its seeds. (Euphorbiaceae)
E. Western poison oak (*Toxicodendron diversilobum*): Shiny leaflets, often with wavy edges, occur in groups of three. Fruits turn from pale yellow to red. (Anacardiaceae)
F. Woolly mullein (*Verbascum thapsus*): Five flower petals; the fruit is a capsule with two chambers. (Scrophulariaceae)

The strategy is to wall off an invader before it can penetrate deeper into a plant's body. Such a defensive reaction is exactly what gall-forming cynipid wasps require to survive. It begins when a female wasp injects her eggs into the living skin of a plant. The eggs hatch, and wormlike wasp larvae stimulate the plant to form a gall around them. The larvae mature into gnat-sized wasps that leave the gall to mate, and the natural cycle repeats. But things do not always go so well

Spiny turban leaf gall

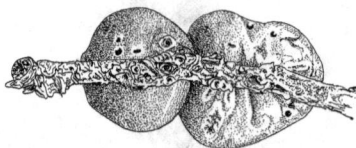

Oak apple stem galls

for the larvae. Various species of flies and moths lay eggs in young galls. Their predatory larvae hatch quickly and the gall soon becomes a war zone of insect larvae pecked at by hungry woodpeckers and jays.

Various creatures cause plant galls to form. **Oak apple galls** can be the size of baseballs, while **spiny turban galls** and other leaf galls are the size of raisins. Size is not the only variable: galls come in an array of shapes, including cones, plates, blobs, and knobs. Cynipid wasps create the galls on oaks that many people are familiar with, but there are various microbes, including fungi and viruses, that also cause these odd plant growths.

Wildflowers and Weeds

Amaranth family (amaranthaceae)

Goosefoot sometimes appears to have whitish dandruff, and their small, greenish flower clusters lack petals. Amaranth

Coyote gourd (*Cucurbita palmata*)
Another means of discouraging predators is odor. Native tradition tells us that Coyote, the mythical trickster of California and much of the West, was wandering the world and discovered this wild melon along his path. He cracked it open and ate the pale yellow flesh inside. He wanted to keep this melon from all other creatures of the world. Consumed with greed, he foolishly urinated on the melon to keep it for himself. To this day, the musty odor of the coyote gourd repels most animals, including coyotes!

is a dusty-looking herb, common in California, that has bluish-green leaves and tight clusters of greenish flowers with three pointed bracts at the base of each flower.

For thousands of years, amaranth helped fuel Mayan civilization; its greenish fruits contain about 50 percent protein, but only wildlife and cattle seem to recognize its value today. Spanish conquerors

Amaranth (*Amaranthus* spp.)

forced Mayans to grow tobacco and various cereal grains in place of this coarse-looking "weed." Stripped of their native diet, which included a variety of seasonal plant foods, Mayans suffered malnutrition. Some evidence suggests that the change in diet was a major factor in their decline. In a similar way, settlers in California spread the seeds of European crops and failed to realize the potential of native foods. California Indians were weakened as their homelands, once rich in nutritious bulbs, herbs, and acorns, were replaced with plowed fields of European row crops.

BORAGE FAMILY

    A.   Cryptantha (*Cryptantha* spp.)
    B.   Fiddleneck (*Amsinckia* spp.)
    C.   Stickseed (*Hackelia* spp.)

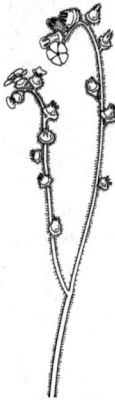 A.
 B.
 C.

BORAGE FAMILY (BORAGINACEAE)

Borage plants have trumpet-shaped flowers crowded along a coiling stem tip. There are four tiny nutlets in the ovary.

Buckwheat Family
Curly dock (*Rumex crispus*)

BUCKWHEAT FAMILY (POLYGONACEAE)

Buckwheat flowers lack petals, but they have four to six sepals. Most buckwheats have a swollen stem with a sheath where leaves arise from the stem. Curly dock (shown here) turns reddish brown by fall.

Buckwheat seeds were ground into flour and cooked by some native people. Various wild species of buckwheat are still abundant and are eaten by animals, and also by people who enjoy brick-heavy but nutritious buckwheat pancakes and biscuits.

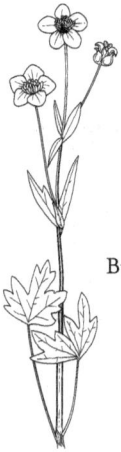

Buttercup Family
Buttercup (*Ranunculus* spp.)

BUTTERCUP FAMILY (RANUNCULACEAE)

Most buttercups have leaves at the base and along the stem. There are usually more than ten stamens, and the petals are silky.

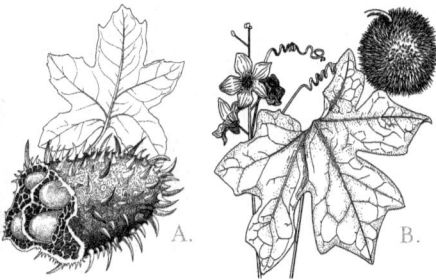

CUCUMBER FAMILY

A. Wild cucumber (*Marah macrocarpa*) (cut away to show seeds inside of fruit)
B. Manroot (*Marah fabacea*)

CUCUMBER FAMILY (CUCURBITACEAE)

A melon fruit develops from beneath the large flower, which is composed of five fused petals. Many species are climbing vines.

Cucumbers guard their seeds with a tough rind containing a matrix of fibers. After the dead fruit has lain on the ground for a few seasons, all that is left is a wad of fiber. Tough but naturally pliable, this "skeleton" makes a good scrub brush to use while soaking in the bathtub.

CARROT FAMILY (APIACEAE)

Plants in this family tend to have umbrella-shaped or flat-topped clusters of tiny flowers,

Carrot Family
Fennel (*Foeniculum vulgare*)

with each flower having five stamens and petals. The ovary is positioned below the petals. Fennel, shown here, was introduced from Europe. Poison hemlock is one species to watch out for in this family because it is extremely toxic. It grows mostly in damp places and has white to yellowish flowers and purple splotches on green stems.

GERANIUM FAMILY (GERANIACEAE)

The flowers of plants in this family have ten stamens and five petals. Geranium fruits are often long and beaklike.

Each of filaree's fruits has a lineup of spear-headed seeds. As the fruit dries out, these spear-seeds coil and drop off the plant. They often catch a ride in animal fur or cotton socks, and eventually they land on the ground. Here, wind helps the coiled filaree seeds to drill downward just enough to plant themselves. Rising from the ground by the thousands, filaree makes its early appearance the next spring.

Geranium Family
Filaree (*Erodium* spp.)

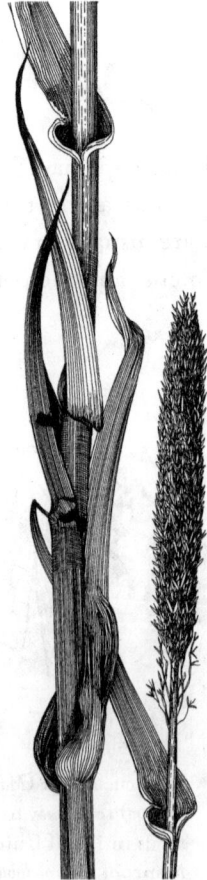

Giant reed (*Arundo donax*)

GRASS FAMILY

A. Wild oats (*Avena fatua*)
B. Foxtail barley (*Hordeum* spp.)
C. Saltgrass (*Distichlis spicata*), a native grass
D. Bermuda grass (*Cynodon* spp.)
E. Watergrass (*Echinochloa* spp.)
F. Canary grass (*Phalaris* spp.)
G. Fescue (*Festuca* spp.)

NATIVE GRASSES

A. Three-awn (*Aristida* spp.)
B. Purple needlegrass (*Nassella pulchra*)
C. Alkali sacaton (*Sporobolus airoides*)
D. Muhly, or deergrass (*Muhlenbergia rigens*)
E. Leymus (*Leymus* spp.)
F. One-sided bluegrass (*Poa secunda*)

GRASS FAMILY (POACEAE)

Most of California's cereal crops are from the grass family. In spring, grass seeds lie at the base of many natural food chains.

Dry grass flowers are enclosed in scale-like lemmas and glumes. Grass flowers are usually elevated to allow pollen to be snatched by breezes. Grass stems are somewhat hollow and often have leaf sheaths.

Grasses grow in length from the middle of the stem or even from beneath the ground. The meristems—growth sites, cells from which new tissue will grow—of most plants are usually limited to the tips of branches or the root. For grasses with meristems scattered throughout their bodies, grazing, mowing, and fire are not life-threatening; as long as some meristem remains, grass can sprout back from a small section of stem. Living meristems allow many grass species to actually multiply when attacked by lawn mowers and weed-eaters.

Visitors to California's interior during the early 1800s brought more than just new cultures. Hitchhiking in the fur and food of introduced livestock were seeds from other lands. Exotic plants flourished here because they had a long history of adapting to livestock in their homeland. The California landscape is now dominated by weedy European grasses that germinate in spring and die all at once after setting seed. Before the time of cattle and plows, interior California was clothed in clumps of perennial bunchgrasses. Thundering herds of pronghorn and elk migrated across this prehistoric landscape. The dominant bunchgrass of California was **purple needlegrass** (*Nassella pulchra*). Its grayish-green clumps produced a backdrop different in color from the springtime green and summer gold of grasslands today. Only about 1 percent of California's natural grasslands remain intact today.

ASPARAGUS FAMILY

A. Blue dicks (*Dichelostemma capitatum*)
B. Ithuriel's spear (*Triteleia laxa*)

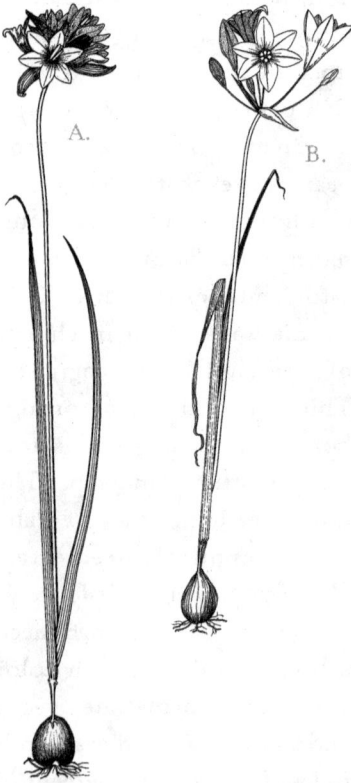

MALLOW FAMILY (MALVACEAE)

The leaves of mallows have a spreading, palmate pattern of veins. Stamens are joined in a central column. Fruits are buttonlike. Most mallows are weedy and have whitish petals. A few medicinal mallows found here originally came from China.

Mallow Family
Common mallow (*Malva neglecta*)

ASPARAGUS FAMILY (ASPARAGACEAE)

The six petals of each flower occur in two petal-like clusters, These perennials (living longer than one year) arise from a starchy underground tuber or rhizome.

MINT FAMILY (LAMIACEAE)

The upper and lower lips of flowers in the mint family are made of five petals fused together. Leaves are opposite and occur at regular levels

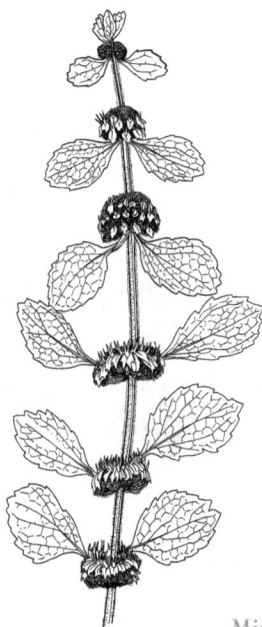

Mint Family
Horehound (*Marrubium vulgare*)

Morning Glory Family
Bindweed (*Convolvulus arvensis*)

along angular stems. Many species have a minty odor.

Native people sometimes wrapped meat in mint leaves, which contain potent natural terpenoids, slowing microbial growth long enough for the meat to dry free of contamination.

MORNING GLORY FAMILY (CONVOLVULACEAE)

The bell-shaped flowers of these plants twist when closing at night. Bindweed, shown here, is a common weed with white and pink flowers.

MUSTARD FAMILY (BRASSICACEAE)

Mustard flowers have only four petals. The fruits are siliques— elongated capsules that split open when dry, with two valves falling away to leave a central partition. They range from being heart-shaped to resembling lumpy bean pods. Wild radish has white to purplish flowers. The other common mustards shown here have yellow petals.

The mustards are another group of exotics imported to California by Europeans. Many were planted by early California missionaries. In time, the cultivated mustards spread and developed a wild flavor that made these naturalized plants unpopular.

MUSTARD FAMILY

A. Wild radish (*Raphanus sativus*)
B. Field mustard (*Brassica rapa*)
C. Black mustard (*Brassica nigra*)

Nightshade family
(solanaceae)

Nightshades are a variable group, but generally they have five petals, and their stamens alternate with the petal tips to form a beak. The fruit is a capsule or berry. Many nightshade species contain toxic alkaloids.

Pea family
(fabaceae)

The lower flower petals in this family are joined into a half-moon shape. There are two wing petals and an upper banner petal as well. The fruits of this family are legumes, resembling pea pods.

Nightshade Family
Tree tobacco (*Nicotiana glauca*)

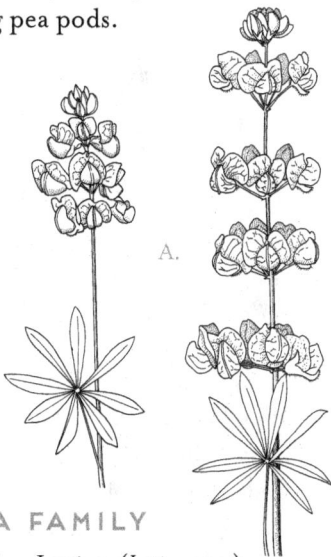

A.

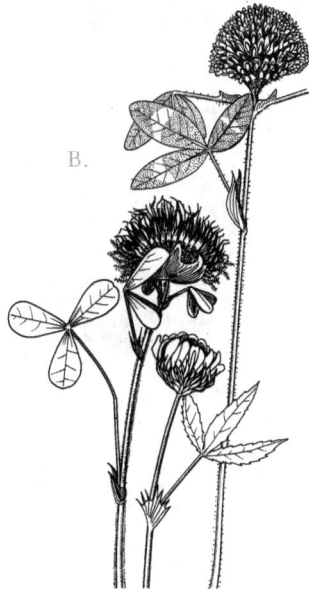

B.

PEA FAMILY
  A. Lupines (*Lupinus* spp.)
  B. Clovers (*Trifolium* spp.)

Plantain family
(PLANTAGINACEAE)

Tiny brown flowers have four petals and sepals and occur as clusters on a leafless stalk. Plantain, shown here, is common in disturbed soil. Nicknamed "white man's foot" by Native Americans, this weed came west along with the cattle that altered native prairies.

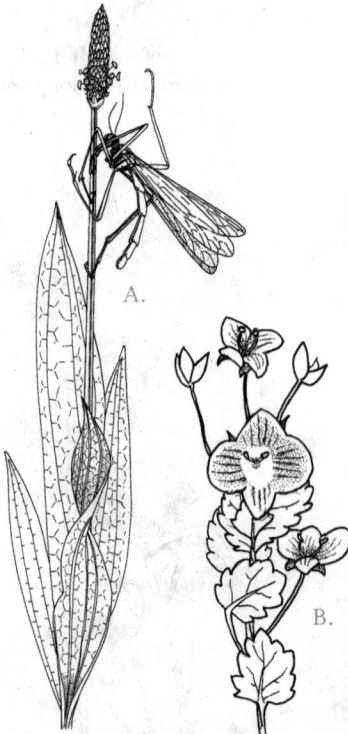

Poppy family
(PAPAVERACEAE)

Four or six silky petals easily drop off the dish-like receptacles of poppies. There are three to four sepals.

Poppy Family
California poppy
(*Eschscholzia californica*)

PLANTAIN FAMILY

A. Plantain (*Plantago* spp.)
B. Speedwell (*Veronica* spp.)

Primrose family
(PRIMULACEAE)

Flowers in the primrose family often have five tiny petals. A long ovary, often bumpy with inner seeds, forms below the flower. Most primroses produce small but colorful wildflowers that bloom in late summer.

Primrose Family
Scarlet pimpernel
(*Anagallis arvensis*)

MINER'S LETTUCE
(MONTIACEAE)

Only two sepals are present on these succulent herbs, below the flower petals. Miner's lettuce, shown here, and several other species store relatively pure water in stems and leaves.

Use caution before using wild lettuce in salads, because these herbs often grow in the shade near trees, a popular location for dogs and coyotes marking their territory with urine.

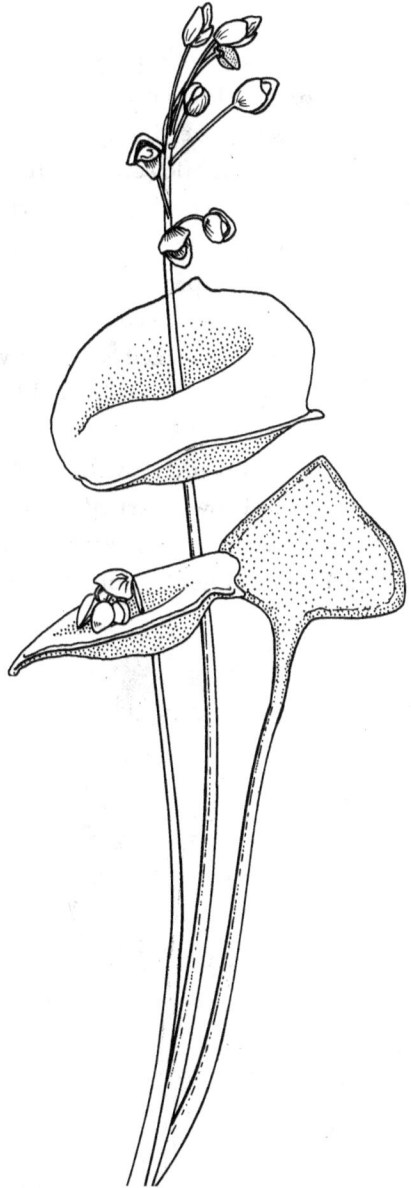

Miner's Lettuce Family
Miner's lettuce (*Claytonia perfoliata*)

Sunflower Family
(Asteraceae)

The most widespread family of California wildflowers, the Asteraceae include various lovely sunflowers as well as the barbed cocklebur, several types of spiny thistle, and many other weedy species.

Sunflowers are composed of tiny flowers packed so closely together that they look like one flower. Disk flowers are crowded together in the middle (the brown part of a garden-variety sunflower) and are surrounded by colorful ray flowers (the golden part). Some sunflowers, such as pineapple weed, lack ray flowers and have only disk flowers. Plants in this family have great reproductive potential because a single sunflower head is loaded with many smaller flowers, each bearing a seed. In some species, a single visit from a pollinating butterfly can result in over fifty fertile seeds!

The sunflowers are well adapted to the heat of Central Valley summers. Horseweed and a host of other fall-blooming sunflowers survive the intense summer sun by

SUNFLOWER FAMILY

A. Gumplant (*Grindelia* spp.)
B. Cocklebur (*Xanthium* spp.)

COMMON YELLOW SUNFLOWERS

A. Pineapple weed (*Matricaria discoidea*)
B. Sunflower (*Helianthus annuus*)
C. Sow thistle (*Sonchus* spp.)
D. Butterweed (*Packera* spp.)
E. Goldfields (*Lasthenia* spp.)

being hairy; a natural growth of hair intercepts sunlight and keeps skin surfaces relatively cool. Gumplants and tarweeds blossom in late summer, a time when the surrounding grasslands are crackling with dead plants. Droplets of tar coat the skin surfaces of these plants and reduce water loss. The presence of these bold blooms is announced by the fragrance of natural tars they produce. With only a few other flowers to chose from in this thirsty landscape, bees, butterflies, and other pollinators flock to the smelly tarweeds and hairy sunflowers of late summer.

WATERLEAF FAMILY
(HYDROPHYLLACEAE)

Flowers are cup-shaped and often have five petals. The stigma, at the top of ovary, is cap-shaped. Stamens are hair-like.

WATERLEAF FAMILY

A. Phacelia (*Phacelia* spp.)
B. Baby blue eyes (*Nemophila* spp.)

FUNGI

Sac Fungi and Lichens | Mushrooms | Eating
Mushrooms | Mushrooms in Grassy Habitats |
Mushrooms near Trees or in Organic Soil |
Some Toxic Fungi | Fungi on Wood

Mushrooms and other types of fungi are often regarded as an unclean menace and associated with death. However, such gloomy outlooks are not realistic. Few fungi of the Valley/Foothill region are harmful to humans, and hundreds of fungal species have economic and ecological benefits.

Mycorrhizae, for example, are fungi that form a partnership with the roots of many plants. The fungal body grows around or even into the root, extending it and allowing it to absorb more water and minerals. In return, the fungus absorbs organic nutrients formed by the plant. Plants typically die or grow poorly without their fungal partners.

Another ecological role of fungi is decomposition. Soil is laced with noodlelike fungi that decompose organic matter. Fungi provide the opportunity for new life by releasing molecular building blocks needed for plant growth. But fungi do not distinguish between a dead log and the wooden foundation of a new home. Among the most destructive fungi are the mildews, blights, and rusts that destroy millions of dollars' worth of plants and property in California each year.

Fungi seem like weird growths that pop up from nowhere and then disappear just as mysteriously. But the visible part of a fungus is just the tip; the rest of the body is quietly spreading, unseen, in lawn, wood, or skin.

Fungal cell walls are mainly composed of chitin, the tough material also found in the skin of insects and shrimp. Microscopic, threadlike cells called hyphae weave together to form the vegetative part of the fungus, the mycelium. Hyphae secrete enzymes that digest fungal food and absorb the dissolved nutrients. When the right conditions occur, hyphae grow into sexual structures that produce spores. Spore color is often used to identify fungi.

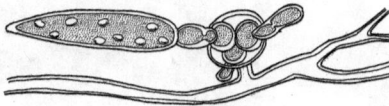

Hyphae (fungal cells)

Spores of various fungi

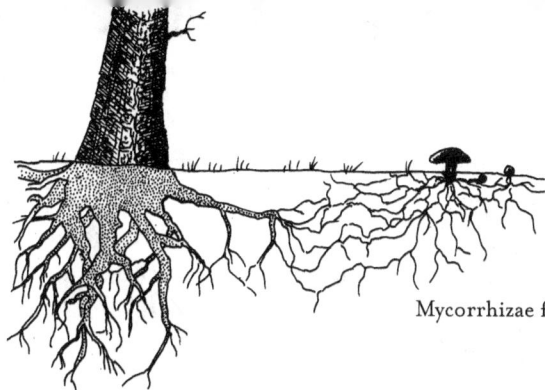

Mycorrhizae fungal connection
with tree roots

Besides sexual reproduction, fungi spread themselves with spores or stolons. A spore released by the parent can germinate and grow into another individual. A single fungus may produce millions of spores in its lifetime, and spores drift nearly everywhere. Stolons are simply strands that branch off the parent fungus, anchor into a food source, and grow into another individual fungus.

Dusting nearly every object on earth is a fine coating of spores produced by microscopic fungi known generally as molds. This odd assortment includes members from various fungal divisions; they have little in common except for being small. Molds produce complex chemicals that help them compete with each other and with bacteria for food. The downside of their important ecological role in decomposition becomes evident when vegetables become slimed and toilet bowls turn green.

Bread mold can contaminate a variety of foods, including bread, fruit, and vegetables. When the packaging of a loaf of bread is opened, the tiny spores that drift about in most homes rush in. Within days, fungal hyphae are penetrating deeply into the loaf. Spores from bread mold can cause chronic pneumonia when growing in the lungs of young children and those with weak immune systems—toss moldy bread before millions of spores from a single loaf begin to float throughout the house. This and related molds can grow in the lungs when inhaled and cause a serious pneumonia.

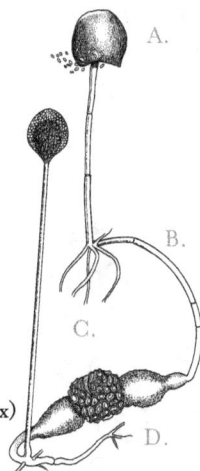

BREAD MOLD (*Rhizopus* spp., magnified 450x)

A. Sporangium
B. Stolon
C. Rhizoid
D. Zygosporangium

Coccidioides immitis is another dangerous mold. Spores of this fungus lie relatively dormant in the dry soil of California's San Joaquin Valley and cause a disease called valley fever. Symptoms are variable; for some people it results in something like the flu, and others may die from it. A variety of other molds cause allergy and weakened immune system disorders. Keep a watchful eye out for molds in your life.

Sac Fungi and Lichens

Sac fungi (division Ascomycota) are rubbery-looking growths that produce spores in tiny sacs. Species of the sac fungus *Penicillum* have been used to extract antibiotics. Various species of *Aspergillus* are used in producing cheese and soft drinks. But many sac fungi are destructive. Yeast infections and diaper rash are caused by a common species of *Candida*. Millions of dollars in crop damage each year are caused by sac fungi that attack fruit trees in California.

The beneficial sac fungi sometimes known as the **lichens** (mostly in division Ascomycota) link up with photosynthetic bacteria or algae. Much of the colorful flaky or crusty stuff growing on rocks in the foothills is composed of a sac fungus that takes its food from algae living within it. The sac fungus is not a parasite on its photosynthetic partner, because there are benefits in the relationship for both; the fungus provides a living mesh where the algae can live. Lichens spend much of their life in a dormant state, losing up to 98 percent of their water during dry times and growing when water is available. Although these crusty growths appear to be tough, they die easily from absorbing airborne metals from polluted air. Absence of lichens in cities is often an indicator of poor air quality.

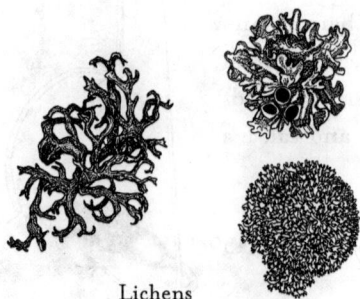

Lichens

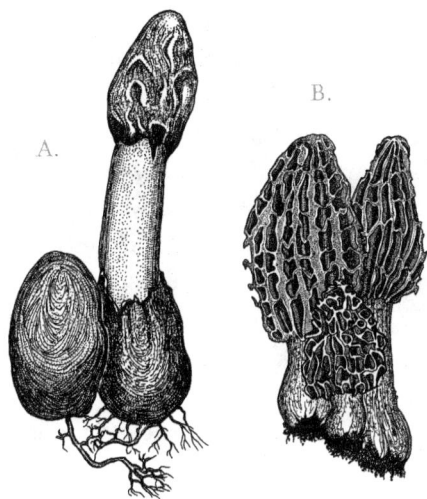

A. Stinkhorn (*Phallus* spp.):
Spongy stalk bolts upward
and is covered in a slimy,
stinking spore mass.

B. Morel (*Morchella* spp.):
Microscopic sacs with 8
spores; folded fleshy cap
and hollow stalk.

Mushrooms
(division Basidiomycota)

From the odd stinkhorn to the deadly amanita, mushrooms come in various shapes, but their similarities are apparent under a microscope. Mushrooms look solid but, like most large fungi, they are actually woven together with layer after layer of thread-like hyphae. All true mushrooms produce spores from tiny clublike arms called basidia. Mushroom gills are packed with basidia, as are the pores of shelf fungi and the inner layers of puffballs.

When the right moisture and temperature conditions occur, the fruiting basidia produce a crop of spores that drift off. One group of mushrooms has the strange trait of dissolving their own gills to disperse spores, resulting in a black, gooey mess. **Stinkhorn** (*Phallus* spp.) mushrooms disperse their spores by producing a stinky, olive-colored slime that sticks to the feet of flies. Some **puffballs** (*Lycoperdon* and *Bovista* spp.) dry out and lie quietly until wind or raindrops compress the sac and force clouds of tiny spores out through the top. The impatient **tumbling puffball** (*Bovista pila*) comes loose from the ground and is blown freely by the wind, dispersing spores as it rolls along—the tumbleweed of the mushroom world. Other fungi protect their spores with toxins or by structures. **Split gill** (*Schizophyllum commune*), a type of shelf fungus, protects its spores by rolling up in dry weather. Some split gills that were placed in a dry tube in 1911 unrolled their gills and released spores when moistened some fifty years later!

Eating Mushrooms

The prospect of eating wild mushrooms is enticing to some people and frightening to others. Some of the most costly foods in the world are fungi; a single matsutake mushroom may bring in as much as $200, and black truffles can go for $500 per pound. On the darker side of this issue is amanita toxin, which has no known antidote.

Toxins are not the only cause of mushroom poisoning. It is often caused by allergic reactions to fungal proteins, or by eating mushrooms that are infested with maggots or bacteria. Eating spoiled mushrooms may seem like odd behavior, but a mushroom in its early stages of decomposition sometimes appears firm and ready for the skillet.

Some mushrooms do produce poisonous compounds. Home testing for mushroom toxins is unreliable and may prove to be deadly: silver coins do not turn black when cooked with all deadly mushrooms; rice does not turn black or red in the presence of toxins; and so it goes with other bits of folklore. Accurate identification is the only way to know if a mushroom is deadly. Fungi have different-looking life stages, making identification tricky for beginners. There is not enough information in this chapter to make certain that a wild mushroom is safe to eat.

In the Valley/Foothill region, the toxin amanitin is found in the **death cap** (*Amanita phalloides*) and the **destroying angel** (*Amanita ocreata*). A single mushroom is enough to cause death. Amanitin is especially dangerous because it cycles in the body for six to twenty-four hours, during which time dramatic symptoms are not necessarily apparent. By the time a person feels sick enough to visit a hospital, the toxin may have already caused irreversible damage to the liver and kidneys. Coprine toxin, which

occurs in **inky cap** mushrooms (*Coprinus atramentarius*), causes an alarming reaction with alcohol in the body. This includes a rapid heartbeat, red ears and nose, and nausea. Recovery is complete. **Psilocybin** toxin (*Psilocybe cubensis*) causes psychedelic color enhancement, visual distortion, and elation, though extreme anxiety may occur.

Effects of some mushroom toxins are somewhat unpredictable. One such case is the **green-spored parasol** (*Chlorophyllum molybdites*). Some people eat this mushroom with no trouble and then recommend it to others who suffer unexpected gastrointestinal upset. Hospitals in Fresno County reported nineteen cases of poisoning in just over two weeks, all the result of this mushroom.

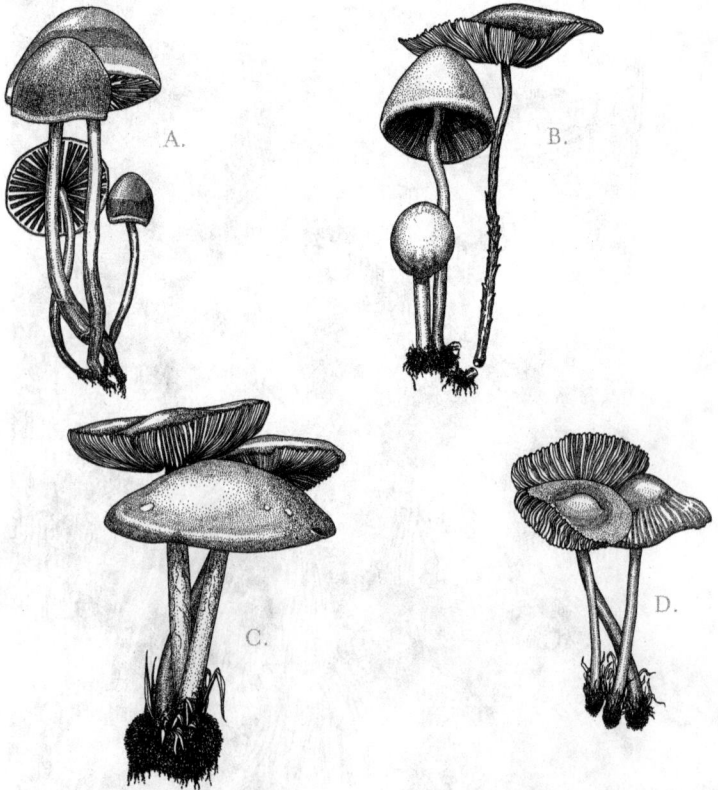

A.

B.

C.

D.

A. Panaeolus (Panaeolus spp.): Common small, brown lawn mushrooms with fragile stalks and black or brown spores.
B. Sunny side up (Bolbitius vitellinus): A fragile mushroom with a radiant yellow cap.
C. Melanoleuca (Melanoleuca spp.): Cap gray to brown; white gills and spores.
D. Fairy ring (Marasmius oreades): Widely spaced gills produce white spores. Often grows in rings on lawns.
E. Horse agaric (Agaricus arvensis): White cap bruises slightly yellow when young; brown spores when mature.
F. Puffballs (Lycoperdon and Bovista spp.): The inside of the spore case is white at first, then turns yellowish, green, or brown with age.
G. Shaggy mane (*Coprinus comatus*): Flaky white pillar ages into a tattered umbrella with black spores.
H. Volvariella (*Volvariella*): Saclike volva at base; pinkish to reddish spores.
I. Cultivated mushroom (*Agaricus bisporus*): Pink to tan gills become dark with brown spores on this robust mushroom.

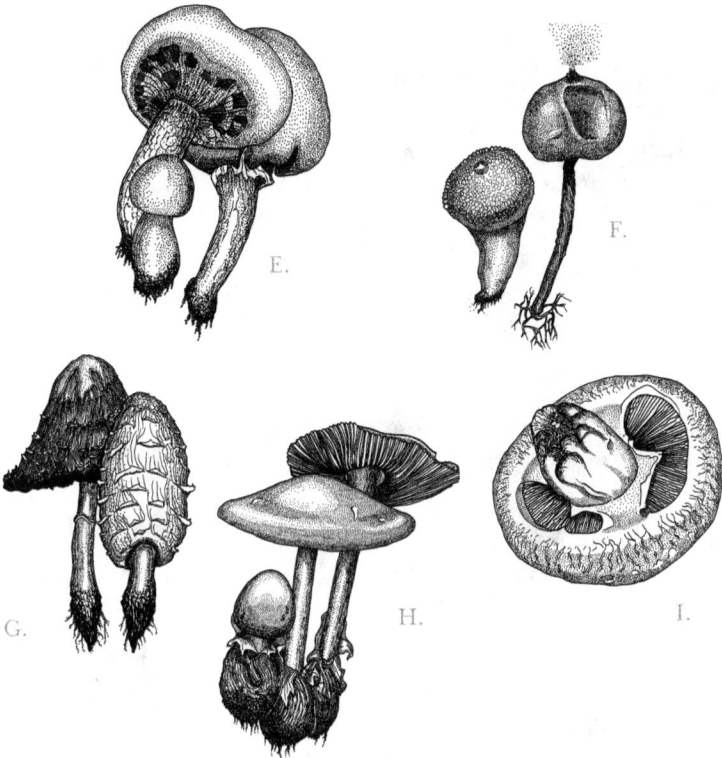

E.

F.

G.

H.

I.

A. Tricholoma (Tricholoma spp.): Many tricholoma species are shades of orange or yellow; spores are white.

B. Destroying angel (*Amanita ocreata*): Swollen base or volva is often underground. These large mushrooms produce white spores. *Amanita ocreata*'s amanitin toxin is deadly.

C. Lentinus (*Lentinus* spp.): White to yellow spores; cap scaly or hairy. One species is known as "train wrecker" because it decomposes railroad ties.

D. Slippery jack (*Suillus* spp.): These robust, thick-stalked mushrooms have slimy caps and white to yellow pores underneath. Spore color includes various shades of brown.

E. Bolete (*Boletus* spp.): Here is a robust, thick-stalked mushroom with olive to brown spores. Pores beneath the cap are generally white, yellow-orange, or brown. Mushroom hunters often stalk this meaty fungus.

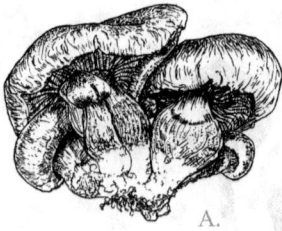

A.

B.

C.

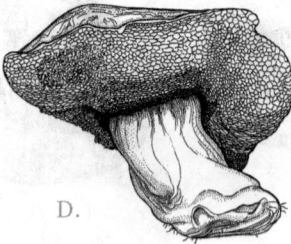

D.

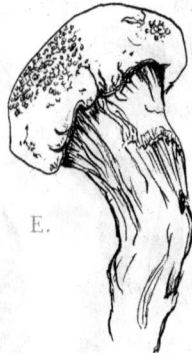

E.

F. Mica cap (*Coprinus micaceus*): Tiny mushrooms with ribbed and flecked caps that release black spores.

G. Shaggy parasol (*Chlorophyllum rhacodes*). Gills change from white to brown with age and release white spores. Cap is broad and often scaly.

H. Inky cap (*Coprinus atramentarius*): Bullet-shaped cap when young; becomes blackened and gooey with black spores when old. Mild toxins tend to react with alcoholic beverages.

I. Magic mushroom (*Psilocybe cubensis*): Stalk bruises blue to green; spore color varies from purple to brownish. Toxic lawn mushrooms are sometimes confused with this illegal toxic species.

J. Deadman's foot (*Pisolithus arrhizus*): Soon after the mushroom emerges, the skin peels off, exposing the smelly and lumpy spore case.

K. Blewit (*Clitocybe nuda*): Purplish cap and gills produce pinkish spores.

A. Death cap (*Amanita phalloides*): Swollen base or saclike volva; universal veil on stalk; white to greenish gills produce white spores. Eating it can result in destruction of the liver and kidneys, and death.

B. Yellow staining agaric (*Agaricus xanthodermus*): Looks like a large grocery store mushroom, but it bruises bright yellow.

C. Green-spored parasol (*Chlorophyllum molybdites*): Olive-colored spores. Greenish, musty-smelling mushroom that causes intense but brief gastrointestinal upset

D. Poison pie (*Hebeloma crustuliniforme*): Crowded pale gills become watery with age and produce brown spores. Toxic but generally not fatal when eaten.

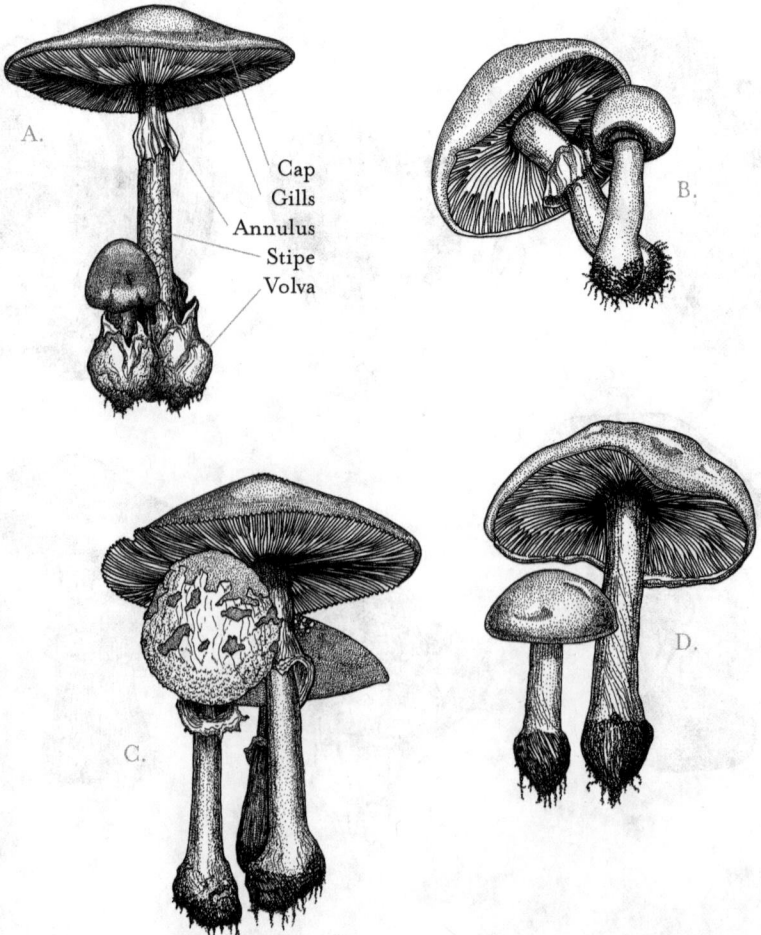

A.

Cap
Gills
Annulus
Stipe
Volva

B.

C.

D.

FUNGI ON WOOD

A. Bird's nest fungus (Cyathus stercoreus): Egglike spore sacs turn dark with age.

B. Oyster mushroom (Pleurotus ostreatus): Tan cap with nutty odor. Pale gills produce white to pale lilac spores.

C. Sulfur shelf (Laetiporus sulphureus): Yellowish cap is smelly and dusty with white spores. Sulfur shelf is sometimes called "chicken of the woods" by people who like to eat it.

D. Comb tooth (Hericium spp.): This fungus develops into what looks like a frozen cascade from the bark of a tree.

E. Turkey tail (Trametes versicolor): Contrasting bands of brown, yellow, cream, and other colors. White to yellowish spores underneath.

F. Artist's conk (Ganoderma spp.): Woody brown cap. White pores underneath produce reddish-brown spores. Undersurface bruises so easily that people sometimes create artwork and messages on these fungi.

G. Earthstars (Geastrum and Astraeus spp.): Outer spore stalk splits apart and exposes inner puffball.

A.

C.

B.

D.

E.

F.

G.

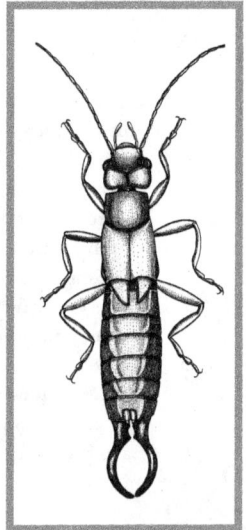

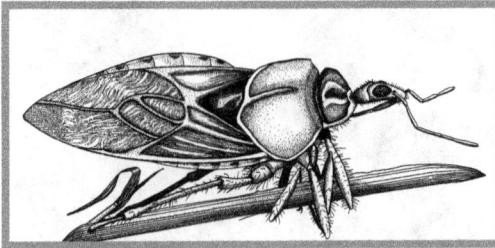

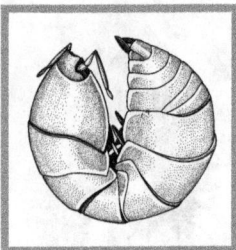

INVERTEBRATES

Sponges, Jellyfish, Moss Animals, and Worms |
Molluscs | Arthropods (Crustaceans, Centipedes,
Millipedes, Scorpions, Solpugids, Mites, Ticks,
Harvestmen, Spiders, Insects)

From squishy sponges to creepy spiders, the trait that all invertebrate animals share is their lack of backbones. Some, such as the jellyfish, are arranged with radial symmetry; they are able to meet their environment from any direction because they have no front or back. Other invertebrates have bilateral symmetry similar to our own, with sensory and brainlike organs close together in a head. Some invertebrates have nervous systems that are netlike or distributed in knots called ganglia, while some are centralized, with a brain and nerve cord.

Another factor in classifying invertebrates is their means of reproduction: some reproduce by division, or budding, in which a clone forms from the base of a parent. Others require courtship and sex, sometimes at great cost, but sex offers a species the advantage of producing genetically unique offspring and thereby, through the generations, adapting to environmental changes.

Sponges, Jellyfish, Moss Animals, and Worms

Following are descriptions of some of the more common or noteworthy invertebrates of the Valley/Foothills region.

Sponges (phylum Porifera) have pores that lead to branching passages lined with collar cells. The whiplike flagella in these collar cells feed the sponge by creating water currents to bring plankton inwards. Not all sponges live in the ocean: freshwater sponges

Freshwater sponge

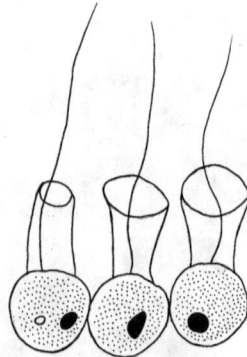

Collar cells of sponge (magnified)

look like globs covered in algae and are easily dismissed as clumps of pond scum.

There are **freshwater jellyfish**, too (phylum Cnidaria). They are a rare but delightful find, as are their relatives the hydras, which attach to the bottom in slow rivers. Jellyfish are famous for their graceful medusa stage and their tentacles. Some tentacles contain entangling threads to capture

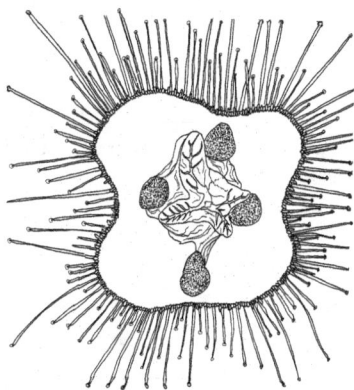

Moss animal colony

moss animals (phylum Bryozoa). They are always colonial, and they secrete a covering for the colony that may be thickly gelatinous or thin and hard. Moss animals use crowns of tentacles to filter feed—to take in water and filter particulate material from it.

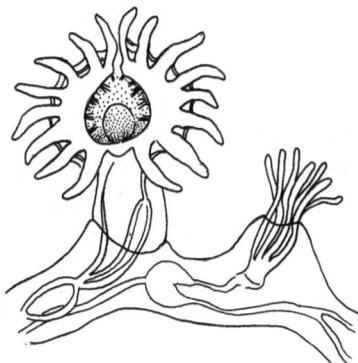

Freshwater jellyfish

prey; others inject toxins that paralyze or kill. An attached stage, called a polyp, forms after a male and female jellyfish mate.

On occasion, in slow waters, you might come across a gelatinous ball the size of a melon. This ball is a colony of tiny

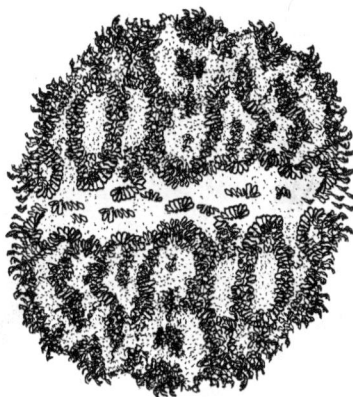

Individual moss animal (magnified)

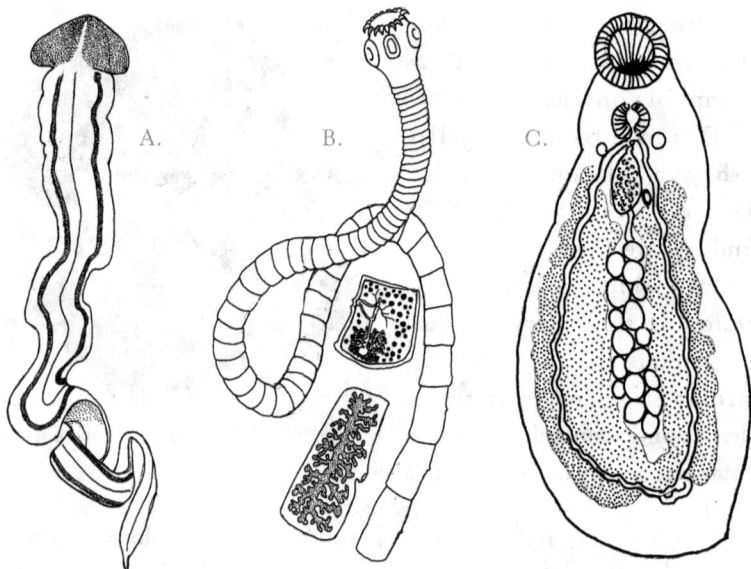

FLATWORMS

    A.   Terrestrial tapeworm. The scolex is a headlike region. In tapeworms it may have suckers and hooks for attachment to the host's intestine.

    B.   Proglottids (reproductive sections of a tapeworm)

    C.   Fluke

Flatworms (phylum Platyhelminthes) are mostly ignored by people. The flatworms of the Valley/Foothill region are scavengers that live in ditches and lawns. A few are fierce parasites. All are flat, and none have a complete digestive tract.

One type of flatworm, the **tapeworm**, consists of a tiny, headlike scolex with hooks and suckers. The rest of the tapeworm is a long chain of proglottids—reproductive sections—full of eggs. Each proglottid contains male and female organs. Mature proglottids detach from the continuous chain and pass out of the host's digestive tract. If an appropriate host eats the eggs, the life cycle is complete.

A common local parasite is the dog tapeworm. Its hosts include dogs and cats that are infected by biting at fleas that harbor immature tapeworms. Beef tapeworm may infect humans who eat meat that hasn't been properly inspected.

The **fluke**, another flatworm, has suckers with which to attach itself to a host, but it lacks the tapeworm's crown of hooks. Human blood flukes infect people in tropical countries where feces enter water; in this region, it is mostly wildlife that suffer fluke infections.

A shovelful of good soil may contain nearly a million **roundworms** (phylum Nematoda) quietly carrying on with their lives. But a few species give this group of tapered worms a bad name. Several **nematodes**

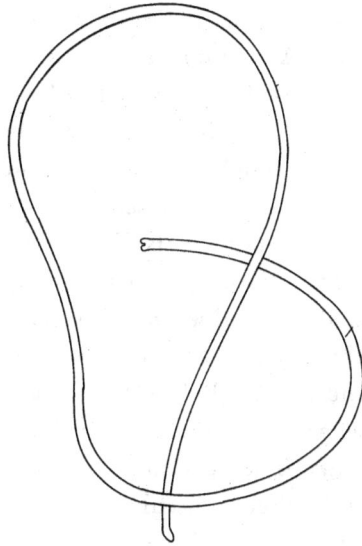

Horsehair worm

attack crops, and others attack animals. **Pinworms** are human parasites common among children (symptoms of this mild infection include fidgeting at night and anal scratching).

The life cycle of the **horsehair worm** (phylum Nematomorpha) begins when adults lay eggs on plants near water. After being eaten, the eggs hatch, and the parasitic worm grows inside an insect's body. When mature, the worm breaks through the insect's body. It does this while the insect is drinking, so that it can move into the water. It's quite a surprise to see a large worm erupt through a cricket.

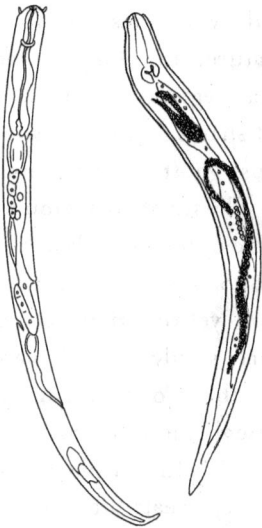

Nematodes (magnified)

In **segmented worms** (phylum Annelida), a body wall composed of segments offers plenty of room for muscle attachment and allows fluid pressure to be raised or lowered in increments. This is why worms can wiggle so amazingly well. **Earthworms** breathe through their moist skin and must seek relief on the surface when rain floods their burrows. They eat soil, grinding it in their muscular gizzards, or forage for leaves and other organic matter. An earthworm has both male and female sex organs in its body but typically mates with another worm: both worms get pregnant.

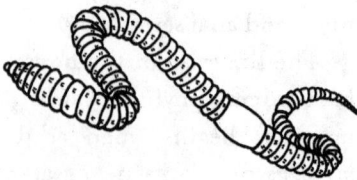

Earthworm

A few bloodsuckers and many tiny predatory leeches, all segmented worms, live in the waterways of the Valley/Foothill region. Bloodsuckers produce anesthetics and anticoagulants that allow them to stealthily make y-shaped wounds and

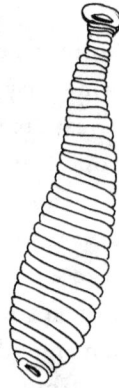

Leech

suck blood from them. **Leeches** have a sucker near their heads and tails with which to attach themselves to their victims, and they can stretch their bodies to a remarkable degree.

When mud flats are exposed during low tide, various creatures give proof of their existence with sudden spouts of water or bubbles from tiny mud volcanoes. One such creature, the **fat innkeeper,** is a segmented spoonworm that shares its burrow with pea crabs, ghost shrimp, and others. The innkeeper's mucus net is one of the marvels of nature: it is so fine that it is invisible to us, yet this tiny net traps a nearly steady flow of plankton that these odd worms pump through their burrows. When the net is full of plankton, the innkeeper eats the whole thing and builds another net.

Molluscs
(phylum Mollusca)

From squishy to rock-hard, all molluscs are similar in having a mantle. This living apron of tissue covers part of the body and helps produce the shell, mostly of calcium carbonate absorbed from water.

Snails and **slugs** feed with a rasplike organ called a radula. They travel about with wavelike contractions of a muscular foot on a path of secreted mucus. Despite their cumbersome shells, snails are accomplished travelers, with some species ranging as far as eighty feet! Snails and slugs have stalklike tentacles. Land snails have eyes on the tips of the longer, second pair of tentacles. The shell provides protection but is not an infallible defense; many creatures feed on snails.

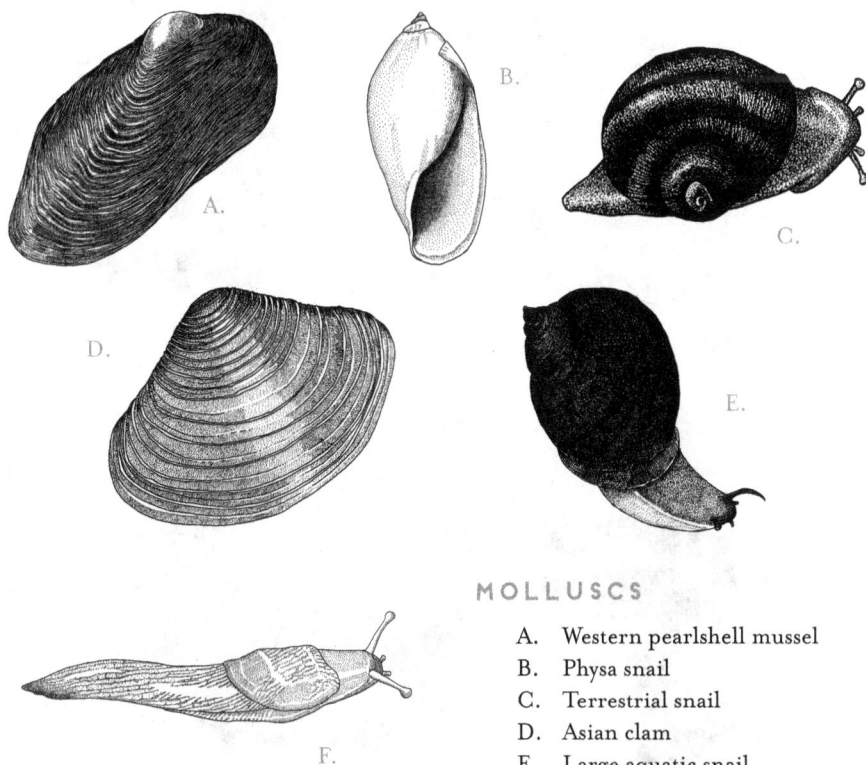

MOLLUSCS

A. Western pearlshell mussel
B. Physa snail
C. Terrestrial snail
D. Asian clam
E. Large aquatic snail
F. Slug

MUDFLAT CREATURES

A. Washington clam
B. Gaper clam
C. Cockle clam
D. Ghost shrimp

E. Proboscis worm
F. Lugworm
G. Fat innkeeper

Clams and mussels, also molluscs, feed by filtering plankton out of water with flaplike gills that they also use for breathing. A single muscular foot is used for burrowing and dragging the shell along. The most widespread and common local bivalve is the invasive **Asian clam** (*Corbicula fluminea*). Native bivalves, such as the **western pearlshell mussels** (*Margaritifera falcata*), were once abundant, some growing to the size of dinner plates. Native mussels were a food source for river otters and, as an anthropologist's report of a Wintun (Sacramento Valley) man emerging from the water with a massive clam in each hand and one clamped between his teeth indicates, for native people. The Wintun people as well as the river otter were almost exterminated in California during the late 1800s.

Arthropods
(phylum Arthropoda)

Tough exoskeletons cover the bodies of arthropods, with muscles positioned inside to control movement. This coat of armor comes with conditions: growth would be impossible inside the confined space if not for the ability of arthropods to molt. They chemically soften and split their exoskeletons, step out of them, and form new ones. As the roots of the word *arthropod* indicate, the jointed legs that allow these armored creatures to move about are common to the phylum.

A spoonful of pond water may have several hundred crustaceans (subphylum Crustacea) paddling about in it. To these tiny shrimp, water is an enormous pasture, but lunch is composed of slimy algae, various microbes, and other shrimp. The exoskeletons of some shrimp are so thin that food can be seen squirting through their guts; other species, such as tadpole shrimp, have an armored carapace. Many crustacean species resemble the shrimplike creatures shown here, but others look nothing like them. Some species of copepod are wormlike, and those that live on fish gills look like space aliens. Even color is variable; some copepods turn bright orange when eating colorful algae.

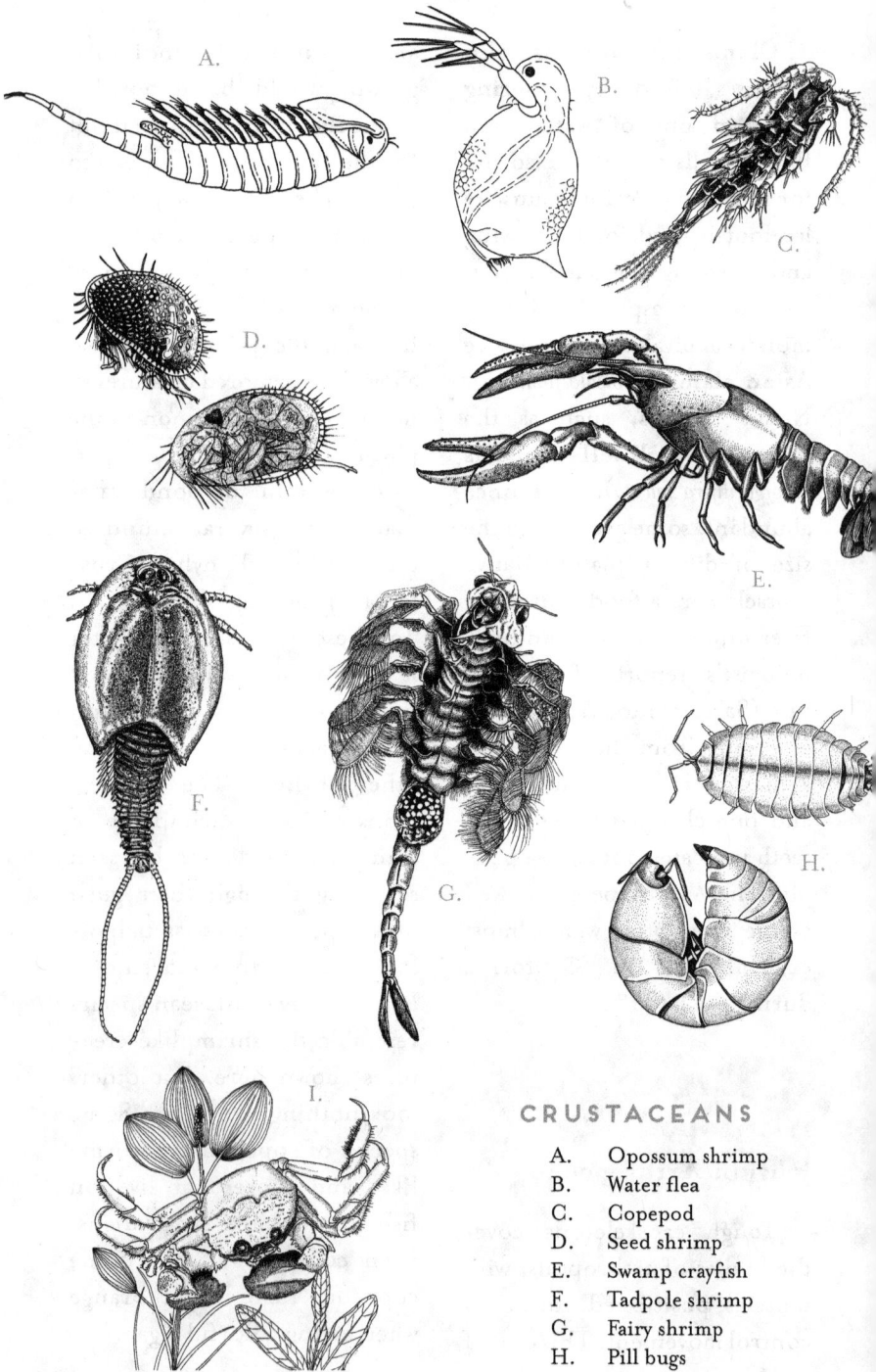

CRUSTACEANS

A. Opossum shrimp
B. Water flea
C. Copepod
D. Seed shrimp
E. Swamp crayfish
F. Tadpole shrimp
G. Fairy shrimp
H. Pill bugs
I. Mitten crab

Wearing the bulky flak jacket known as an exoskeleton drives large crustaceans to inhabit water, where they can fortify their armor with calcium carbonate. Add to this the fact that crustaceans breathe through gills instead of lungs, and it is amazing that they ever leave the water. But some do. Pill bugs must stay in humid locations or risk death due to drying of their gills. They also roll to keep their gills moist and to protect themselves from curious children. **Chinese mitten crabs** (*Eriocheir sinensis*) invaded the San Francisco estuary and moved inland through the Sacramento–San Joaquin River Delta in 1996. There are reports of the crabs wandering streets at night and entering houses or swimming pools. Crayfish leave water to search for food. They are scavengers but will attack nearly anything they can catch.

Centipedes (subphylum Myriapoda) run down their prey at night or under the cover of boards, using claw-like poisonous legs to inject a paralyzing poison. Centipedes have one pair of legs per trunk segment. Young hatchlings have fewer trunk segments than adults of their species, adding more at each molt until maturity. Different species of centipede can have from fifteen to over one hundred trunk segments, with an average of perhaps thirty-five. Some species have appendages near their tails that mimic antennae, causing confusion in their enemies.

Sleepy-looking **millipedes** (also myriapods) are sometimes discovered cruising around under boards or out at night, grazing on dead plant matter. Millipedes have two pairs of

Millipede

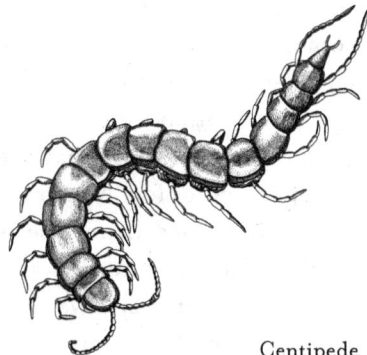

Centipede

legs per trunk segment and a domelike back, and they often curl up into spirals. Common Valley/Foothill species are an inch or less in length, though a robust species four inches in length occurs in the foothills. Millipedes lack the poison claws of centipedes, but most have glands that produce smelly chemicals for defense. Most animals avoid eating the sluggish and toothless millipede because of these bitter chemicals. Some millipedes produce arsenic, which has harmed children who consumed these interesting crawlers.

Arachnids
(class Arachnida)

Arachnids are the class of arthropods that includes scorpions, mites, ticks, spiders, and others that possess the fearsome, fanglike or pincerlike cheliceræ—appendages near the mouth often modified for grasping or piercing—that the group is famous for. Arachnids are distinguished from insects by their lack of antennae and by the fact that they have only two major body sections, one of which carries four pairs of walking legs.

The abdomen of a **scorpion** (order Scorpiones) is modified at its tip into a tail bearing a stinger. No California scorpion inflicts a sting lethal to humans; the potency of their neurotoxin varies with the species. The scorpion uses its large pincers in concert with the stinger to grab and subdue prey, and the tiny pincers on its chelicerae to ingest food. Scorpions hunt a variety of small prey at night. Young scorpions are born alive and may ride on their mother's back for several days.

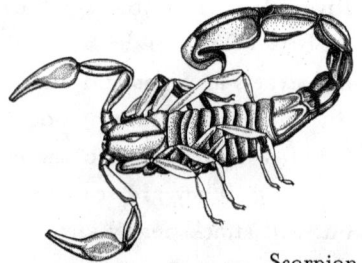

Scorpion

Solpugids (order Solifugae) are also known as sun scorpions or wind scorpions. They live in deserts and grasslands, sometimes causing a stir when they enter houses. Being without venom, they can be welcome,

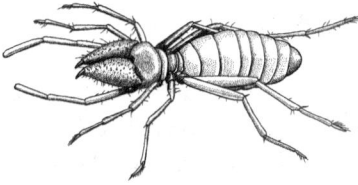
Solpugid

pest-killing guests. They look ferocious and, although no poison glands have been discovered on these animals, this appearance is not misleading. Their large chelicerae are modified into efficient killing pincers that macerate their invertebrate prey. The long pedipalps, easily mistaken for a fifth pair of legs, are sensory appendages.

Mites and **ticks** (superorder Acari) are a strange group of tiny creepers that occur nearly everywhere. Both are arthropods with a single body region. Most have eight legs as adults. Look no farther than your own eyebrows to find the nearest mite. Parasitic mites live in the skin of animals, and

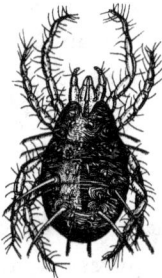
Tick

other mites are scavengers that feed on bits of food and even dead flakes of skin. Mites attack crops, and dust mites cause a lot of allergies. Fortunately for us, there are predatory mites with the heroic job of hunting down and eating other mites.

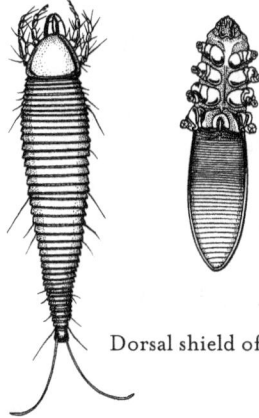
Dorsal shield of a mite

Ticks are parasites that suck blood. They can survive for months without a meal. Some species wait on plants for a passing host, which they grab onto with their legs. If you are bitten, use tweezers to grasp the tick as close to your skin as possible, then pull steadily. Save the tick in alcohol so it can be identified. If it is a **western blacklegged tick** (*Ixodes pacificus*), there is a possibility of Lyme disease. This disease is caused by the bacterium *Borrelia burgdorferi*, transmitted by the bites of infected ticks. Pronunciation is not the only challenge this tiny spirochete presents: early symptoms of Lyme disease are variable and unreliable. A reddish rash around the bite occurs in less than 50 percent of cases; many people experience nothing at first. Unfortunately, only in the later stages of infection do recognizable symptoms such as nervous system disorders occur. If you are bitten, remove the tick and save it for examination.

Any tick can cause infection, especially if fragments of its head are not removed from the skin. Wash the area with soap and water, and consult a doctor if it turns red or looks infected. After all this talk of danger, it gives some peace of mind to know that relatively few adult ticks are infected in our area, though a higher risk appears to be from tick nymphs.

Harvestmen (daddy longlegs, order Opiliones) are jolly-looking arachnids that cruise around without webs. In spite of their looks, harvestmen are not spiders; their two body sections are attached over a broad area unlike the pinched-in midriffs of true spiders. Feeding through a long tube, they eat plant juices, decaying organic matter, and, of course, aphids. The harvestman's legs fall off easily, which distracts enemies long enough to save the individual's life at the cost of a leg. Lost legs are not regenerated.

Harvestmen

How common are **spiders**? There is probably one of these arachnids (order Araneae) less than seven feet away from you right now. It may have killed a young cockroach or mosquito last night, so let it carry on with the business of life. Regardless of the variety of colors and shapes they may come in, all true spiders, like scorpions, ticks, harvestmen, and the other arachnids, have chelicerae (poisonous fangs), four pairs of walking legs, and two body sections. Spider-web silk has greater tensile strength than steel. Accurate identification often requires a magnifying

Pedipalps (feeding legs)

Chelicerae (fangs)

Cephalothorax
(head and trunk body region)

Four pairs of walking legs

Abdomen (hind body region)

Spinnerets

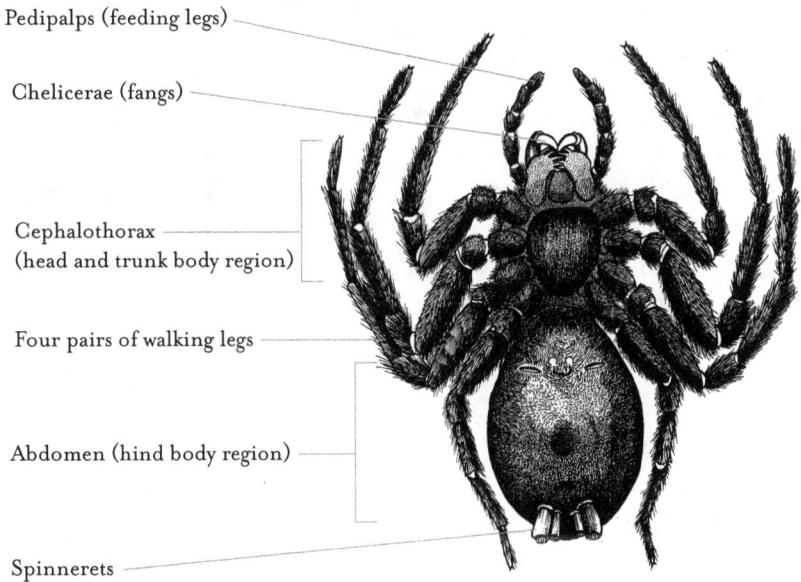

glass because the tiny claws, eyes, and other features are critical.

Only two spider species of the Valley/Foothill region are dangerous to humans. Doctors report that most of the sores that people blame on spiders are in reality caused by tiny splinters, plant bristles, and other non-spider sources. One of the two local spiders with enough fang and venom to hurt us is the female **black widow** (*Latrodectus* spp., a comb-footed spider, family Theridiidae). The shiny black

female has a round abdomen with a red blotch—shapes vary— on her ventral (belly) surface. The smaller male, which does not bite humans, has an elongated abdomen with white and red on its sides. The female may kill and eat the male after mating, the habit which earned the species its somber common name. Black widow webs are commonly found in sheltered locations, including garages, sheds, grapevines, and woodpiles. The web has a funnel-like retreat with a white egg sac

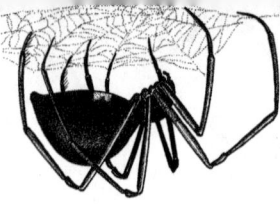
Black widow

Young black widow
(spiderling)

House spider

nearby. Unlike the handiwork of most spiders, the black widow's thick web makes a crackling sound when it is torn. Black widow spiderlings are patterned orange, brown, and white, darkening after each molt.

Comb-footed spiders, which also include the **American house spider** (*Parasteatoda tepidariorum*), among others, are named for the bristles on their hind legs. Most have long legs and a globular abdomen. The main web is often a sloppy-looking affair, but these spiders use their leg bristles to "comb" and wrap their prey into silken cocoons.

The other dangerous spider of this region—though the question of its presence in California is still controversial—is the **brown recluse** (*Loxosceles reclusa*). The members of its family, Sicariidae, are called violin spiders because of the fiddle-shaped pattern on their cephalothorax. Recluse venom is a digestive protein that dissolves flesh, leaving a hole

where bacterial infections cause greater damage than the original venom. The fiddle pattern on its back, three rows of two eyes, and a crease in the hind part of the cephalothorax help identify the rare brown recluse.

The **crab spider** (family Thomisidae) earns its name by scuttling like a crab. It waits with "arms" outstretched among the flowers or leaves of

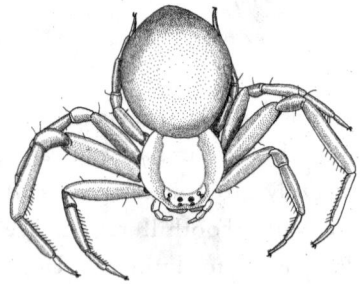
Crab spider

Brown recluse

a plant, ambushing its unsuspecting prey with a quick forward snap of its long, powerful limbs. These spiders do not spin webs to catch prey, although the males may temporarily tie down their mates with silk webbing. Crab spiders have eight eyes in two rows, with many of the eyes raised on bumps.

Ground spider

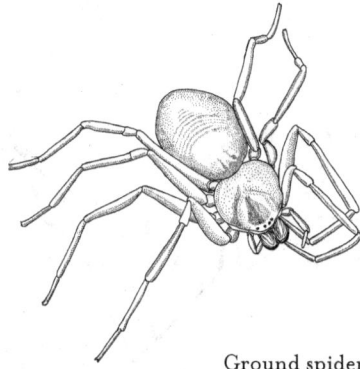

Wolf spider

Wolf spiders (family Lycosidae) are hairy spiders that run down their prey in a wolf-like manner. These are the harmless (to humans) spiders that are often seen dashing for cover around homes. Most are nocturnal and have a reflective layer in their eyes that creates an "eyeshine." Some dwell in vertical, silk-lined burrows with tubes protruding above the surface; others rove on the ground. Young spiderlings may ride on their mother's back for several days.

Ground spiders (family Gnaphosidae) behave very much like wolf spiders but are less often seen because their favorite haunts are among stones and leaf litter. Ground spiders spin tube-shaped nests that protect them from enemies and are rarely used to trap prey.

Tunnel-shaped webs under rocks or logs, or cobwebs in rolled-up leaves, are typical signs of **sac spiders**, also called **running spiders** (family Clubionidae). Many species resemble ants, a feature that may cause predators to

Sac spider (running spider)

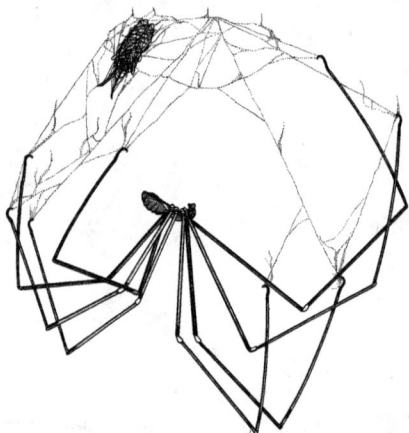
Cellar spider

ignore them. This is a difficult group to identify, because their appearance varies. The sac web, sometimes an antlike appearance, and two tiny claws with bristles between them on each foot are distinguishing features.

Morning dew often gathers on the flat webs of **funnel weavers** (family Agelenidae) strung on grass or along fences. The spider hides in a silken tunnel to one side of the main web. When a flying insect

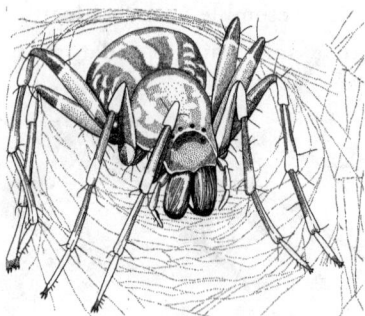
Funnel weaver

hits the upper tunnel, it falls into a silk sheet below, where it is greeted by the spider's fangs. The features that separate this group from sac spiders are a hind pair of spinnerets that are twice the length of the first set, and three tiny claws on the tip of each foot.

Cellar spiders (family Pholcidae) are common residents outside buildings, where they spin their webs near porch lights and under eaves. Young cellar spiders often disperse indoors and are the most common source of cobwebs inside homes. Webs are irregular and woven of fairly weak silk. These spiders use their long legs to wrap silk around small insects and other spiders caught in their webs. Cellar spiders rock from side to side, vibrating their webs, when threatened.

Rapidly scuttling forward or sideways with ease, **jumping spiders** (family Salticidae) stalk prey and then often capture it with an impressive leap. This is accomplished with a hydraulic, rather than muscular, mechanism and with the help of an anchor line of silk that spins out as they jump, allowing them to regain their former

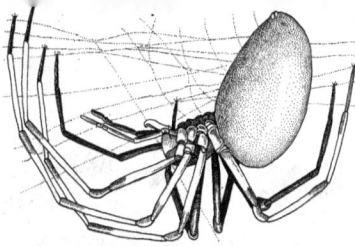

Long-jawed orb weavers

position. Their high blood pressure facilitates this unusual jumping method. Of the spiders, they have the sharpest vision. Jumping spiders have eight eyes (as do most spiders), with two larger eyes pointing forward. This array makes it difficult to surprise one of these small hunters. Jumping spiders can be found on tree trunks, sides of houses, window screens, and the ground. Silk-lined shelters may well indicate their presence.

Few natural sights are as scary as the thumb-sized, black and yellow, bulbous body of some local **orb weaver** species (family Araneidae). They do no harm to humans but are the bane of insects. Orb weavers build webs with support lines that radiate outward from the middle. They cross these support webs with sticky, circular hoops, and they can rebuild the entire web each night. A radiating web, two rows of equal-sized eyes, and a bump or knob on the outer part of the jaws help identify orb weavers.

Long-jawed orb weavers (family Tetragnathidae) are leggy spiders that build webs shaped like wagon wheels. Their jaws are so long they look like legs. Long-jawed weavers tend to spin webs at night, trapping flying insects. It is frightening to run into one of their webs, but these spiders are harmless to us.

Jumping spider

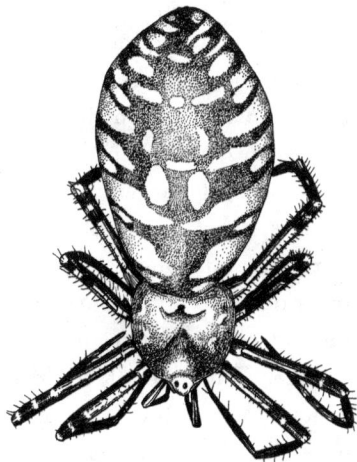

Orb weavers

Trapdoor spider

Trapdoor spiders (family Ctenizidae) are large, black arachnids that live in and around their silk-lined burrows and retreat rather than bite when threatened. The spider holds the hinged lid of its trap shut until it senses the vibrations made by passing prey. Then it suddenly rushes out to drag its victim back into the burrow.

Tarantulas (family Theraphosidae), as many people know, are big and hairy. These spiders, too, live in silk-lined burrows. Although imposing in appearance, most are docile and retreat when given the chance. The bite of a North American tarantula is about as potent as a bee sting. Tarantulas are most active at night, when they seek insect prey. Males do not live very long, and do not molt even once after maturity. They wander, often on cloudy days, seeking females with whom to mate. Females have lived as long as twenty-five years in captivity. The body of a tarantula is covered with fine hairs that can irritate human skin.

Tarantula

Insects
(class Insecta)

Creeping, soaring, and digging, insects are nearly everywhere. Insects, too, are arthropods, and they reign supreme in terms of sheer numbers. Their success comes in part from wings, formed from extensions of the exoskeleton; flight is a handy mechanism when the present moment is unsatisfactory or lethal. Metamorphosis is another key to their success, allowing many insects to maximize food resources by eating different foods at different stages of development. Varied types of mouthparts and body shapes also contribute. Just as there is a tool for every job, there is an insect lifestyle that fits somewhere in most food chains on Earth.

Metamorphosis means change, and it is the stuff of fairy tales and nightmares. Some creepy larvae become graceful butterflies, and some harmless larvae become bloodsuckers. In gradual metamorphosis, the nymph usually looks like a small version of the adult. In complete metamorphosis, the larva is often wormlike and looks nothing like the adult. The larva feeds, molting several times, until it is plump enough to enter the pupa stage. As a pupa, the insect does not move about, but this is a time of great change, as wings and other adult structures develop. The adult insect then emerges, and the crumpled wings are pumped full of blood, expanded, and allowed to dry.

A close look at insects shows that they are beautifully designed. Each of the three body sections is composed of specialized performance parts. The head serves as the sensory center, and it ingests food. The mouthparts, far more than just lips, help determine how an insect makes a living, and they also help with identification. Two antennae are present and serve as organs of touch, smell, or hearing. Compound eyes, made of many sub-units, enable insects to be incredibly alert with very little brain power. The middle body part, the thorax, is specialized for locomotion, bearing all three pairs of legs and the wings. Many insects have an eardrum, called the tympanum, on the thorax also. Finally, the abdomen serves for digestion and reproduction. Adult insects breathe through a network of tubes with openings, called spiracles, on the sides of the body.

GRADUAL METAMORPHOSIS

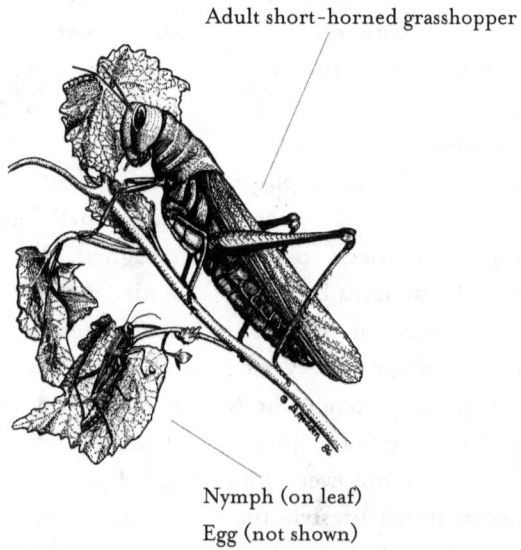

Adult short-horned grasshopper

Nymph (on leaf)
Egg (not shown)

COMPLETE METAMORPHOSIS

A. Pupa
B. Adult ladybird beetle
C. Larva

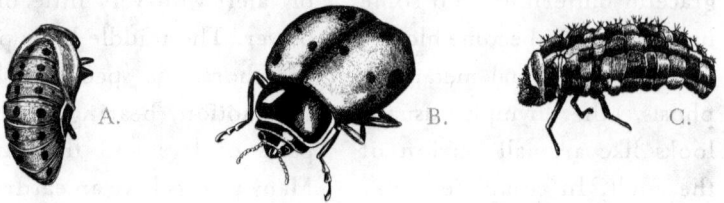

A.   B.   C.

INSECT ANATOMY

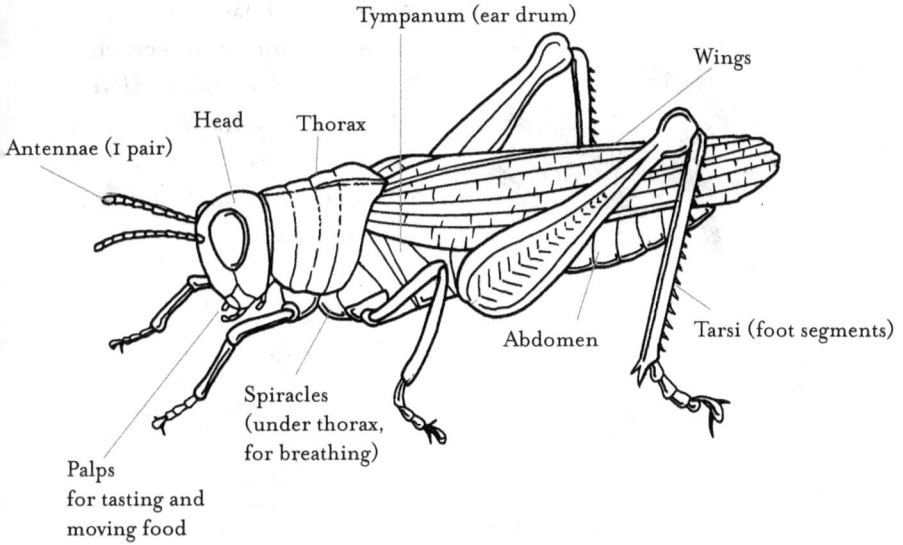

Tympanum (ear drum)

Wings

Head

Thorax

Antennae (1 pair)

Tarsi (foot segments)

Abdomen

Spiracles
(under thorax,
for breathing)

Palps
for tasting and
moving food

INSECT MOUTHPARTS

    A.   Chewing, such as found in beetles and grasshoppers
    B.   Slicing-sucking, such as found in some flies
    C.   Piercing-sucking, such as found in bugs, lice, fleas, and some flies
    D.   Spongelike, such as found in houseflies
    E.   Chewing and lapping, such as found in bees
    F.   Tubular, such as found in butterflies and moths

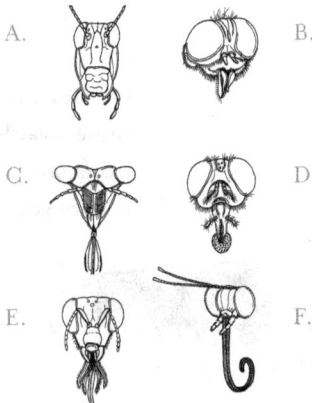

A.

B.

C.

D.

E.

F.

Springtails
(Order Collembola)

Springtails

Springtails look like tiny jumping seeds and are rarely noticed until a clump of them shows up on a damp sidewalk. Most feed upon decaying plants. The name "springtail" refers to their means of locomotion: a furcula—an appendage rising from one of the abdominal segments—provides these insects with the ability to spring twenty times their own body length.

Silverfish
(Order Zygentoma)

Wingless but fast on six legs, these streamlined insects chew on a variety of foods. **Silverfish** eat all sorts of starches, including those found in book bindings, curtains, and even wallpaper paste.

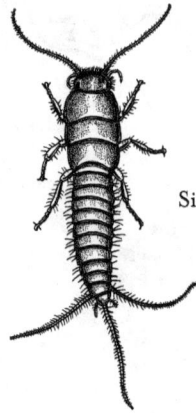

Silverfish

Rockhoppers
(Order Archaeognatha)

Rockhoppers make their living outdoors, eating decaying matter. Two long antennae on the head and three long bristles on the abdomen allow these crawlers to sense vibrations, and whisk—they are gone before you ever see them.

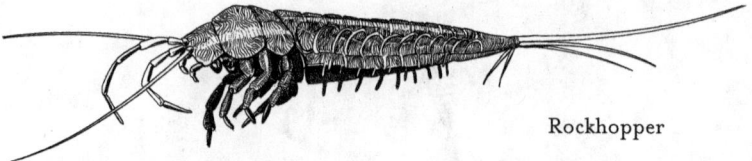

Rockhopper

Dragonflies and Damselflies
(ORDER ODONATA)

Dragonflies are masters of the sky, outflying most other insects. Their massive eyes allow them to locate flying prey, which they capture in baskets formed by their bristled legs. Sometimes they eat while still in flight. Male and female dragonflies are often seen in tandem, mating in the air. Dragonflies are territorial, investigating intruders, including humans, and driving away competition.

Dragonflies are just one of the odd assortment of insects that spend most of their lives as submerged larvae, lying along a river bottom and avoiding predators with the help of camouflage coloration and secretive lifestyles. Most adult dragonflies live for only a few weeks. Dragonfly nymphs are ideally adapted to aquatic life. They breathe through the rectum, which can also squirt out a jet of water to propel them away from enemies. Their lower lips are hinged and have grasping structures on the end so they can reach out and grab unsuspect-ing prey, including tadpoles, aquatic insects, and minnows.

Damselflies can be distinguished from dragonflies by the fact that their wings are folded back along their bodies when at rest.

Caddisfly larvae (order Trichoptera), also aquatic, build cases of sand or bark and feed primarily on algae, with some species scavenging. The larvae of web-spinning caddisflies live in retreats of debris and spin silken nets to capture drifting prey.

Mayflies
(ORDER EPHEMEROPTERA)

Mayflies are delicate, aquatic insects. The adults lack functional mouthparts and often live less than one day, dying soon after breeding. They are

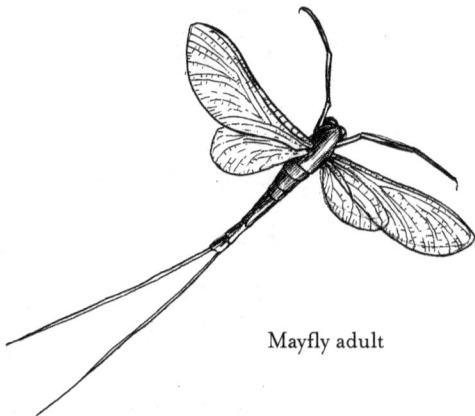

Mayfly adult

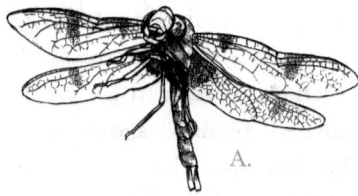

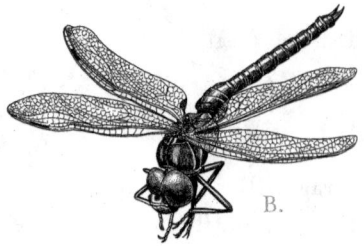

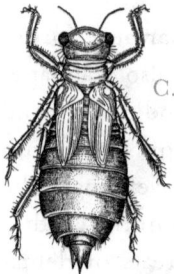

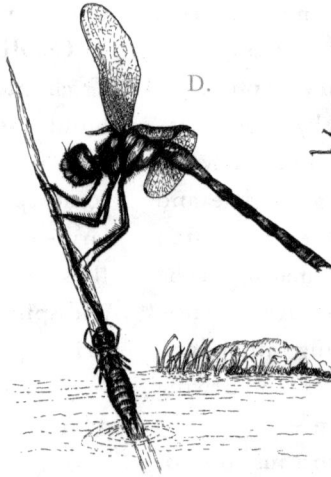

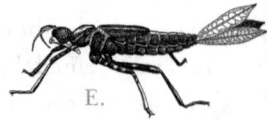

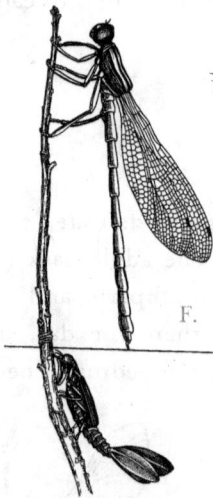

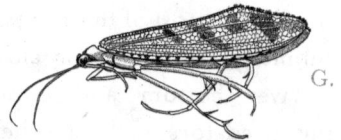

DRAGONFLIES AND DAMSELFLIES

A. Skimmer
B. Darner
C. Dragonfly nymph
D. Dragonfly and nymph

E. Damselfly nymph
F. Damselfly and nymph
G. Caddisfly adult
H. Caddisfly larva

Mayfly larva

in color with brightly colored second wings visible only in flight. The brownish "tobacco" that a grasshopper spits out is a defense; it consists of distasteful blood from thin-walled blood vessels in its mouth.

weak fliers, found in greatest numbers near the body of water in which they hatch, though waifs are often blown far away. Mayflies are important links in many aquatic food chains because they are abundant and because they transform plants and bottom ooze into the insect flesh that many fishes eat.

GRASSHOPPERS AND THEIR RELATIVES (ORDER ORTHOPTERA)

These hoppers are known for their songs and for their ability to leap, although many species do neither very well. All have a gradual metamorphosis; the nymph looks like a small version of the adult.

Some **grasshoppers** (family Acrididae) make a loud, crackling noise when flying, caused by the wings striking each other. Most are gray or brown

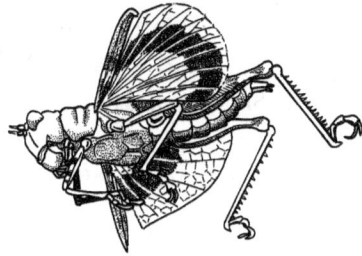
Grasshopper

Katydids (family Tettigoniidae) are often green, from the pigments in the leaves they eat. These insects are nocturnal and may be heard from trees on summer nights. Some emit a quick "lisp" by rubbing their wings together. A relative, the cone-headed grasshopper, makes a loud buzzing sound at night.

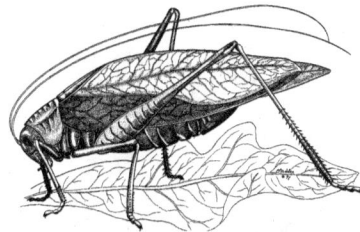
Katydid

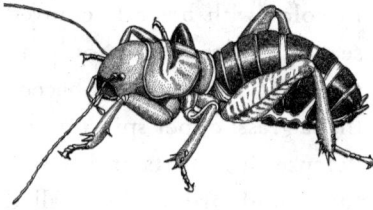
Jerusalem cricket

Also known as potato bug, or *niña de la tierra* (child of the earth), the **Jerusalem cricket** (family Stenopelmatidae) takes refuge under rocks and in burrows. At night this large and strangely hideous creature may forage for plants and insects on the surface, where it often startles those humans it encounters. It is not venomous, although it may bite if roughly handled.

Unlike this quiet species, **field crickets** (family Gryllidae) are accomplished songsters. The males rub their wings together to produce sound in order to attract females.

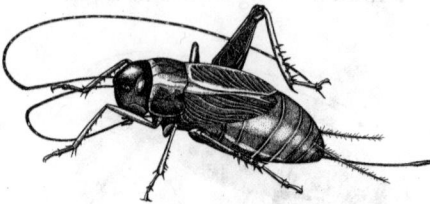
Cricket

Mantids are unusual among insects in their ability to rotate their heads, an adaptation that allows them to locate prey while lying in wait, otherwise motionless. The front legs, on the elongated thorax, are modified—flattened and spiny—to grasp prey. They are often called "praying mantids" because they wait for prey with

Mantid

front legs raised as though in prayer. Adults often appear at porch lights. The female may decapitate the male during mating and eat him afterward. This odd act is explained by the amount of energy required to produce eggs. Mantids overwinter as eggs in their distinctive egg cases. After hatching, the young may eat siblings that hatch later. The large, bright-green mantids most commonly seen are an introduced species.

Cockroaches inspire fear and loathing in many people. They are pests that invade our homes, but they are more annoying than harmful. Their Latin name refers to their aversion to light, and they do scuttle for cover in its presence. Metamorphosis is gradual, with wingless nymphs hatching from egg cases. The waxy layer covering the body prevents water loss and is the source of an unpleasant odor associated with cockroaches. The largest cockroach in the Valley/Foothill region is the **American cockroach** (*Peripla-neta americana*). Both sexes are reddish-brown in color and their wings extend past the ends of their abdomens. The dark and shiny **oriental cockroaches** (*Blatta orientalis*) are the most common pest cockroach

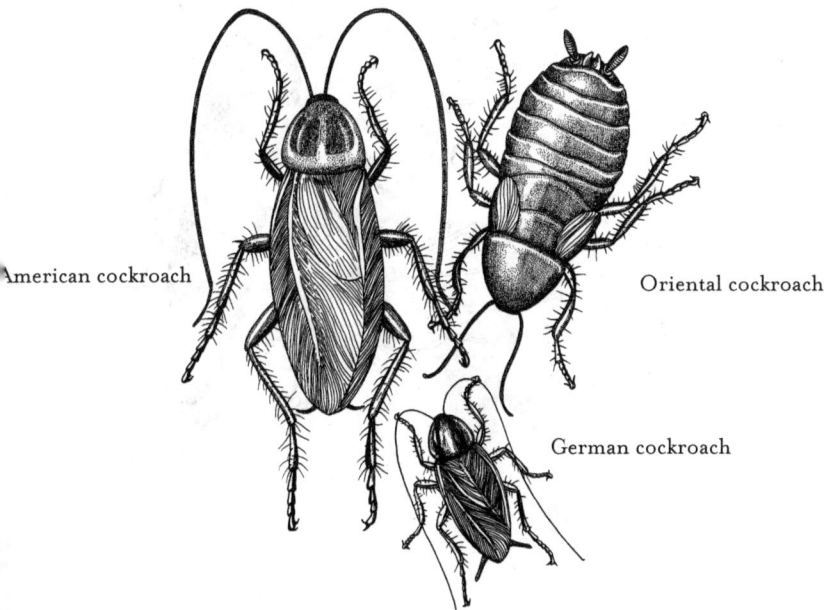

American cockroach

Oriental cockroach

German cockroach

in the region. They are called water bugs because they thrive in the presence of moisture. Males have wings shorter than their abdomens and females have small, functionless wings.

Termites depend on their social structure to provide temperature-regulated colonies where they can survive. Each colony includes several castes. The worker caste are wingless males and females. The larger (and fewer) soldier caste defends the colony. Some soldiers have such big jaws that they cannot feed themselves and are fed by workers. Reproductives fly off to establish new colonies: a single queen is accompanied by a male, the king, who periodically fertilizes her.

Termites lack the enzymes necessary to digest wood and depend upon microbes living in their guts to do this for them. Such microbes are passed from older to younger termites by mutual feeding. Of the fifteen or so termite species in California, about five are pests to humans.

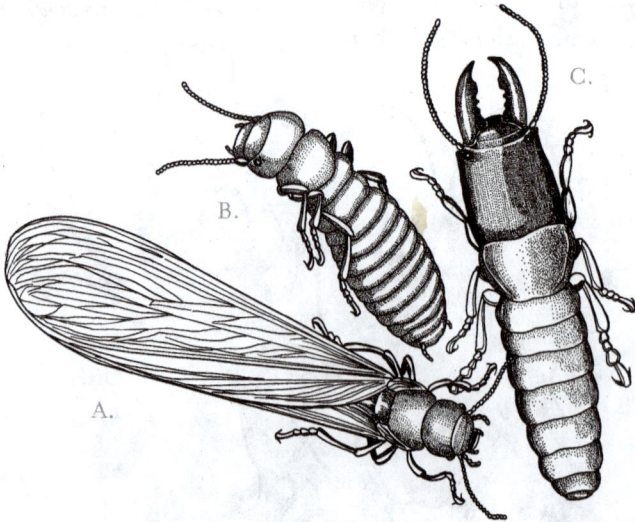

TERMITE CASTES

   A. Winged reproductive
   B. Worker
   C. Soldier

EARWIGS
(ORDER DERMAPTERA)

These insects are named earwigs because of the belief that they enter the ears of humans, an event more likely to happen to people who sleep on straw than to anyone else. Most earwigs are nocturnal and are only seen when rocks or bricks are turned over, or when water in our yards floods them out of cracks. They are mostly scavengers. The pincers are used for defense and can deliver a good pinch.

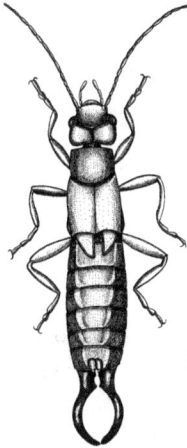

Earwig

THRIPS
(ORDER THYSANOPTERA)

Thrips are tiny insects rarely noticed by the general public, but they are considerable pests to some farmers and gardeners. They have four extremely narrow wings edged with long, hairlike fringe. To find thrips, search flowers for many tiny specks. Distinctively edged wings and the behavior of arching upward just before taking flight identify them as thrips. Most thrips feed by sucking liquid from plants.

Thrips

LICE (ORDER PSOCODEA)

Most biting lice parasitize birds and will eat feathers, hair, or outer skin of the host; they do not attack humans. Sucking lice, on the other hand, have tubelike mouthparts for sucking blood. Their eggs, attached to host hairs, are known as nits. Two types are specific to humans: the **head louse** (*Pediculus humanus*), or "cootie," and the **crab louse** (*Pthirus pubis*). Lice have been responsible for the spread of epidemic typhus and much death and suffering. Among conditions frequently transmitted by schoolchildren, head lice are second only to the common cold.

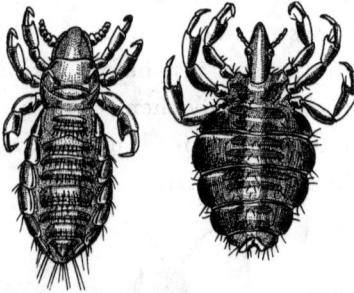

Lice

FLEAS
(ORDER SIPHONAPTERA)

Fleas are adapted to live as parasites on larger hosts, where they feed on blood. Their larvae look like white worms with long hairs. The **cat flea** (*Ctenocephalides felis*), which is the most common flea of domestic cats and dogs, is itself a host in the life cycle of a parasitic tapeworm. The **ground squirrel**

Flea

flea (*Diamanus montanus*), which parasitizes the California ground squirrel, is known to carry bubonic plague, which is transmitted by the bite of infected fleas. Deer mice carry fleas that may transmit hantavirus, which can cause a fatal illness.

BUGS
(ORDER HEMIPTERA)

Bugs are insects, but not all insects are bugs. The hard forewings of a true bug form an "X" across its back. Bug mouthparts, designed for piercing and sucking, are used to feed on plants and to prey on or parasitize others. Many bugs use glands to produce smelly substances that repel enemies or attract mates.

Assassin bugs (family Reduviidae) stalk over plants or wait by flowers for passing prey. Their forelegs are spiny, for holding their victims, whom they inject with venom

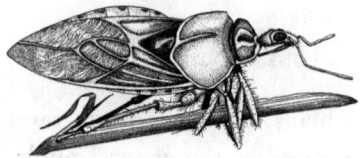

Assassin bug

BUGS

A. Bordered plant bugs
B. Shield bug
C. Squash bugs
D. Toad bug
E. Leaf-footed bug
F. Stink bug
G. Box elder bug

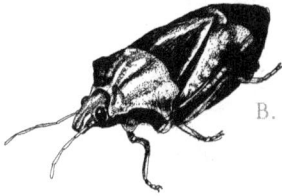

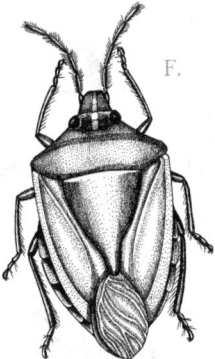

A.

B.

C.

D.

E.

F.

G.

from their short beaks. The venom not only kills the prey but liquefies it so the assassin bug can suck up its meal. The bite is reported to be quite painful to humans.

Backswimmers (family Notonectidae) carry a bubble of air for breathing when submerged. They eat other insects and even small fishes and can inflict a painful bite. Backswimmers often rest at the surface of the water, floating head down with their long, oarlike legs extended. Backswimmers are good fliers and can end up in swimming pools and other bodies of water. During courtship, some male backswimmers produce sound by rubbing their beaks with their front legs.

Water boatmen (family Corixidae), another water-dwelling species, rarely bite. Most feed on algae, small aquatic organisms, or decaying matter.

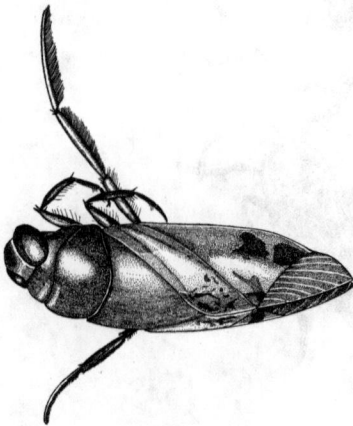

Water boatmen

Water striders (family Gerridae) have tiny, wax-coated hairs on their feet that prevent them from breaking through the surface film of water as they skate. This arrangement, together with their long legs, allows them to literally walk on water. They use their middle pair of legs to row about, and the hind legs help distribute their weight and balance them. The front legs are used to capture and hold prey. These

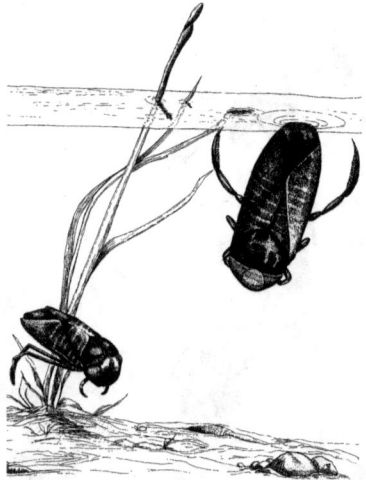

Backswimmer

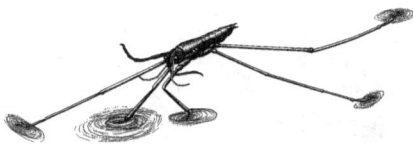

Water strider

bugs skate around quiet areas, attacking insects that fall into the water.

CICADAS, APHIDS, LEAF-HOPPERS, AND SPITTLEBUGS (ORDER HEMIPTERA)

Members of this group all have piercing-sucking mouthparts, which they use to attack their plant food.

Each species of **cicada** (family Cicadidae) makes a unique, high-pitched buzzing sound that attracts a mate. Cicadas spend the majority of their life underground as nymphs feeding on roots. One eastern species requires seventeen years to complete its life cycle; two to five years is typical of our California species.

Honeydew is a characteristic sign of **aphids** (family Aphididae). It consists of plant sap passed out of the anus of the aphid. Aphids can transfer plant diseases with their piercing mouthparts.

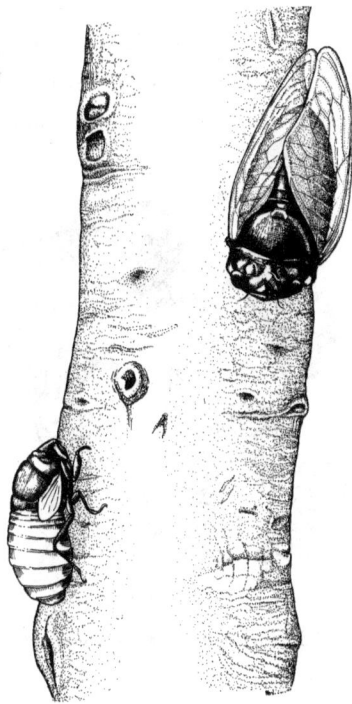

Cicada larva and winged adult

Leafhoppers (family Cicadellidae) also secrete honeydew. These are the tiny hoppers that are almost always present during the growing season in vineyards and backyards.

Leafhopper

A frothy hiding spot that resembles spit is made by larval **spittlebugs** (family Aphrophoridae).

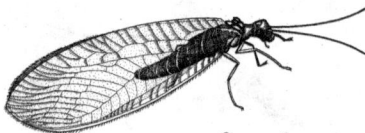

Green lacewing

Brown lacewing

gauzy, iridescent wings. Some adults feed on the honeydew of aphids, an ideal situation because young lacewings are voracious predators of aphids. In fact, immature green lacewings are called aphid lions. Green lacewings lay their eggs on stalks to stave off predators and to prevent the first hatchling from consuming its siblings! These insects frequent porch lights and produce a noxious odor when disturbed.

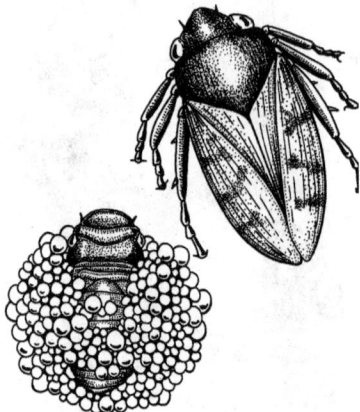

Spittlebug larva and winged adult

LACEWINGS AND ANT LIONS (ORDER NEUROPTERA)

Adults of this order may be seen fluttering around at night on their four long wings, but it is at the larval stage that most of the action happens. These effective assassins kill by ambushing their prey. Some of the lacewing larvae go so far as to tangle debris or even the dry carcasses of their prey on their body hairs to provide camouflage.

Green lacewings (family Chrysopidae) are a beautiful pale green, with shiny, copper-colored eyes and

The larvae of **ant lions** (family Myrmeleontidae) use cone-shaped pitfall traps dug

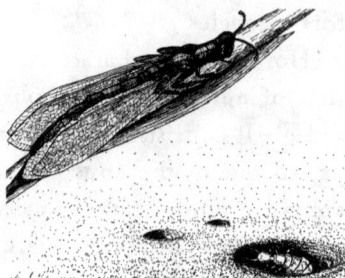

Ant lion adult
(its larva is in the sand pit below)

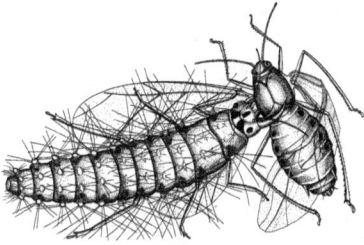

Lacewing larva attacking an aphid

in loose sand to capture their prey. The larva waits at the bottom of the pit, completely buried except for its head and large jaws. When an ant or other small insect stumbles into the pit, the larva showers it with sand, preventing its escape, and then grabs it and injects it with venom. These larvae are sometimes called doodlebugs, perhaps because of the doodle-like tracks they leave when they move, typically backward, on the surface. Adults have very small mouthparts; many do not feed and those that do usually eat flower nectar.

BEETLES

(ORDER COLLEOPTERA)

The beetle's shell-like outer wings, called elytra, form a straight line where they meet. When not in use, the inner wings fold under the elytra.

Both adults and grubs have chewing mouthparts.

Most species of predacious **ground beetle** (family Carabidae) are dark-colored and long-legged with dark wing covers often etched with parallel lines. Some species are flightless. These beetles are often nocturnal predators, taking refuge by daylight under logs, rocks, or other debris.

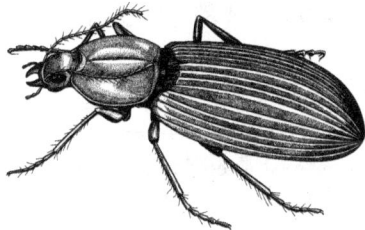

Ground beetle

Tiger beetles (family Cicindelidae) are dark, active, predatory beetles that fly in short bursts along the ground as they pursue their prey. These beetles are often mistaken for flies in their typical sandy beach

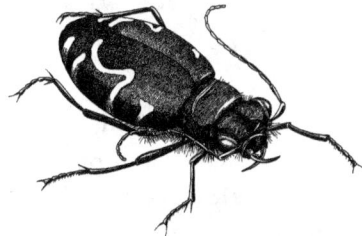

Tiger beetle

habitat, where they may hunt true flies.

Many people think of ancient Egypt when **scarab beetles** (family Scarabaeidae) are mentioned, because that culture placed so much importance on this beetle. Scarabs can be identified, in part, by their heavy, roundish bodies and distinctive antennae, the last segments of which are flattened into fanlike plates. These beetles feed on plant material, dung, or carrion. Most species make a hissing sound when held, and they buzz in flight. The c-shaped larvae eat roots; people often dig them up while gardening.

Scarab beetle and larva

Although generally nondescript in color and of modest size, **click beetles** (family Elateridae) are very distinctive.

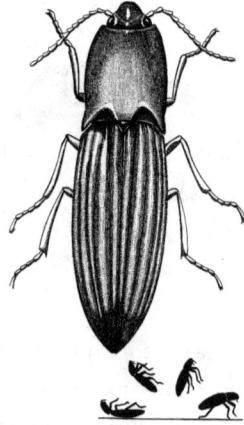

Click beetle

When turned on its back, a click beetle can click a spine on the underside of its thorax, providing enough spring so that the beetle flips up, often landing on its feet. This ability makes this beetle a favorite of children, who seem to never tire of its neat trick.

The antennae for which **long-horned beetles** (family Cerambycidae) are named are

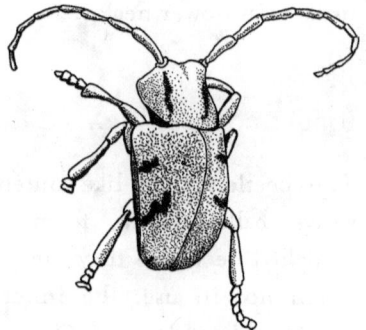

Long-horned beetles

at least two-thirds the length of their bodies. These beetles spend much of their lives as larvae in tunnels and galleries that they carve into the trunks of trees and the roots of shrubs. The grub stage of most takes a year, but larvae have been known to emerge as beetles after living for ten years or more in wooden furniture.

Weevils (family Curculionidae) have long snouts with small, biting mouthparts on the end. Most of California's weevil species—more than a thousand—feign death by folding their legs and dropping from their perches when threatened.

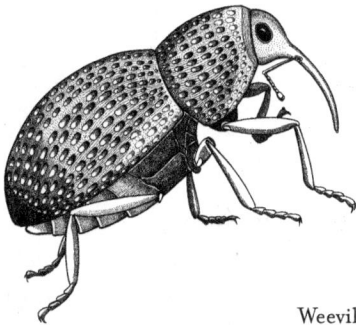

Weevil

This behavior can be used to help control the **plum curculio** (*Conotrachelus nenuphar*): sheets are placed under infested trees, banged on to induce the faint, and then gathered up with the beetles and destroyed.

There are more than one hundred and twenty-five species of **ladybird beetle** (ladybugs, family Coccinellidae) in California, but the most familiar is probably the **convergent ladybird beetle** (*Hippodamia convergens*), named for two converging white lines that are easily mistaken for eyes. These are the ladybugs that migrate and gather in large clumps in the Sierra Nevada, where they overwinter. The larvae of ladybugs have long, tapered, alligator-like bodies. Adults and larvae of this species feed primarily on aphids.

Some species of **darkling beetle** (family Tenebrionidae) enter homes when the weather cools; others, like the flour beetle, are year-round pests of stored grains. Among the larger of the darkling beetles are the **stink beetles** (*Eleodes* spp.), which often wander in the open. A stink beetle lifts its

Darkling beetle

abdomen to signal its distaste-fulness to predators, some-times simultaneously releasing bad-smelling chemicals.

The shiny, black **scavenger water beetle** (family Hydro-philidae) is the submarine of the insect world. Its body is streamlined, with grooves for the legs and numerous tiny hairs that enable the beetle to trap a bubble of air for breath-ing underwater. The back legs are flattened, with many bris-tles, and can function as pad-dles. The weird-looking lar-vae of most aquatic beetles are predators.

Scavenger water beetles

FLIES (ORDER DIPTERA)

The mouthparts of flies are a wonder of adaptation. They include the piercing, sucking mouthparts of mosquitoes, the sponging mouthparts of houseflies, and the cutting mouthparts of horseflies. The second pair of a fly's wings is reduced to knobs that function like gyroscopes to keep the fly balanced while it changes flight direction.

Crane fly

Crane flies (family Tipuli-dae) are common and harmless inhabitants of Valley/Foothill homes. They resemble mos-quitoes, except they are not bloodsuckers. Larger species are referred to as "mosquito hawks" and are falsely believed to eat mosquitoes: most adults do not feed, and larvae eat underground roots.

Mosquitoes (family Culicidae) transmit many diseases that cause misery and death. Several species of mosquito are known to complete their life cycle in less than ten days. Some have been found breeding in old tires, flower vases, or even water-filled footprints. The larva is the "wiggler," named for its swimming motion. Only the female mosquito feeds on blood; the male eats flower nectar. Not all species eat blood, and many of the bloodsucking species do not feed on humans.

Hover flies (family Syrphidae), named for their habit of hovering as though suspended from a thread, look like bees or wasps—a resemblance that serves to protect these stingless flies. Hover fly larvae often feed on insect pests, including aphids. The adults pollinate flowers as they feed.

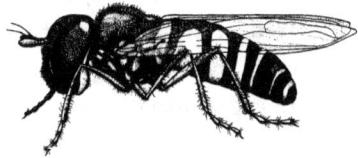

Hover fly

Dangerous-looking **robber flies** (family Asilidae) are efficient predators on other insects. Their mouthparts are adapted for piercing and are used to kill prey with a stab to the thorax. The predatory larvae live in rotting logs. When hunting, an adult robber fly sits on a favorite perch, surveying the area with its watchful eyes until another insect flies by. It then springs into flight from its perch, something like a falcon. If successful in the hunt, it carries its victim to a perch and devours it.

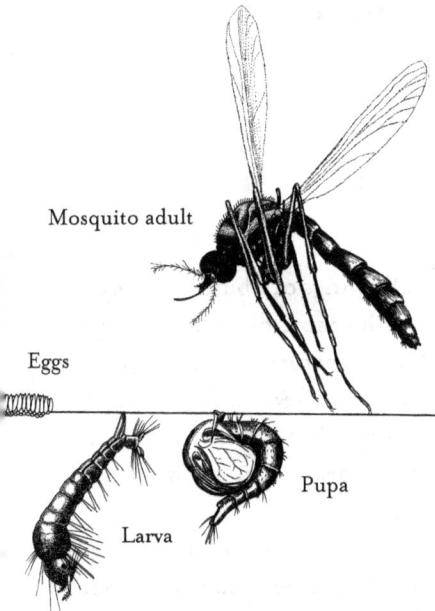

Mosquito adult

Eggs

Larva

Pupa

Robber fly

Some species of **housefly** (family Muscidae) bite, but the majority do not. They are considered pests by virtue of their role in spreading disease and their aggravating habit of trying to consume perspiration from our skin and moisture from our eyes. Their eggs, laid in decaying material or excrement, hatch into maggots. Exposed to the germs associated with such materials, adult flies can transmit diseases. The housefly is believed to be dependent on humans for its life, as it is found only in association with humans.

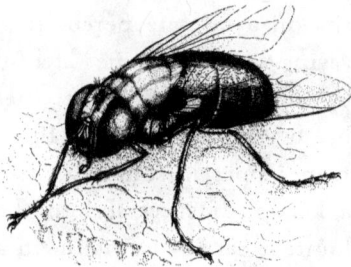

Housefly

Tachinid flies (family Tachinidae) resemble houseflies, but these bristly insects lay eggs on caterpillars. After the eggs hatch, the fly larvae burrow into the host's body, where they feed until ready to pupate, at which time they leave the host. Often the host dies, literally eaten alive from the inside out!

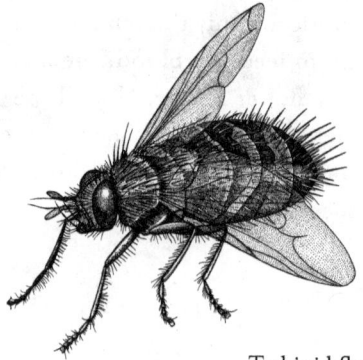

Tachinid fly

Butterflies and Moths (order Lepidoptera)

The wings of a butterfly are one of the most beautiful sights in all of nature. Their apparent fragility serves a function: the scales fall off easily, which aids their bearers in escaping from spider webs.

Butterflies are brightly colored day fliers and have clubbed antennae. Moths, on the other hand, are usually active at night and have threadlike, comblike, or plumelike antennae.

Besides carpet beetles, **clothes moths** (family Tineidae) are the insects most likely to ruin your favorite sweater. The adults are rather plain and avoid light, retreating to dark closets. The larvae form cases from spun silk and bits of their food, creating a natural camouflage.

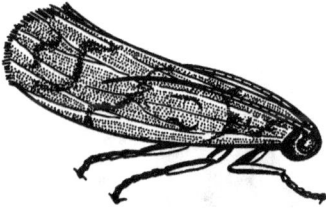

Clothes moth

The brownish **snout moth** (family Pyralidae) often invades our homes to eat stored, dry foods. One in particular, the **Mediterranean flour moth** (*Ephestia kuehniella*), commonly infests homes. It seeks dark cupboards or pantries in which to lay its eggs. In order to eliminate these pests, it is necessary to remove infested foods and seal off the replacements. The **Indian meal moth** (*Plodia interpunctella*) has similar habits but has white-and-rust-colored wings.

There are over one thousand species of **cutworm moth** (family Noctuidae) in California. The adults are typically robust, brownish moths, and the larvae are thick caterpillars with cryptic coloration. The larvae of many species hide in grass by day and emerge at night to feed.

Cutworm moth

Snout moth

Sphinx moths on white datura flowers

Sphinx moths (family Sphingidae) are most active after dusk, when they flit from blossom to blossom, hovering like hummingbirds as they collect nectar with tubular mouthparts that may be three times the length of their large bodies (wingspan can be as large as five and a half inches). Larvae have hornlike projections and are called hornworms.

Skippers (family Hesperiidae) of the Valley/Foothill region are most often orange or yellow, although there are other color patterns. The dull-colored caterpillars feed inside leaf shelters they form by fastening the edges of a leaf together with silk.

Fluttering and gliding through towns each summer are the colorful **brush-footed butterflies** (family Nymphalidae). The red admiral, painted lady, mourning cloak, and others add splashes of color to the sky on warm days. The front pair of legs is reduced and covered by long hairs, earning this family its name. The brushes are used to sample nectar.

The **mourning cloak** (*Nymphalis antiopa*) is one of the first butterflies to appear in

Mourning cloak

spring, because it is one of the few butterflies that overwinter as adults rather than as an eggs or pupae. It was named for its dark wings, thought to resemble the clothing of widows in mourning. The wings are edged by a row of small, blue dots and a contrasting band of creamy yellow. Larvae feed heavily in groups on willow and poplar trees.

The black forewing of the **red admiral** (*Vanessa atalanta*) has a transverse band of orangish-

Red admiral

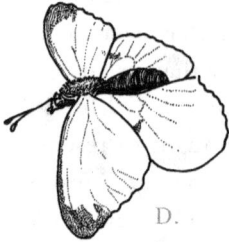

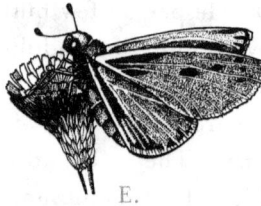

COMMON BUTTERFLIES

A. Blues
B. Fritillary
C. Tortoise shell
D. Sarah orange-tips
E. Skipper
F. Checkerspot
G. Angelwing
H. Anise swallowtail
I. Buckeye

red and white spots. Larvae feed on stinging nettles, which grow along lowland rivers. Each larva produces its own private dining hall by fastening the edges of a leaf together with silk.

The striking orange, black, and white wing pattern of the **painted lady** (*Vanessa cardui*) is not visible when the wings are folded; rather a brownish, cryptic pattern is displayed. The larvae form a feeding "nest" by connecting two or more leaves with silk. Every several years, large northward movements of painted lady populations are reported in California. The steady spring-time parade of butterflies fluttering through towns and fields goes on for several weeks.

An orange spot near the tip of the black forewing and a white slash through the middle of both wings helps identify the **California sister** (*Adelpha*

Lorquin's admiral

californica). It is typical of oak woodlands, where larvae feed on oaks and adults feed on wildflower nectar. **Lorquin's admiral** (*Limenitis lorquini*) looks similar, though the entire outer edge of the forewing is orange. The larvae, which resemble bird droppings, feed on willow.

The larvae of **monarchs** (*Danaus plexippus*) feed on milk-weed. The toxins in milkweed accumulate in all life stages of the monarch and can cause vomiting and irregular heart-beat in predators—those who have once eaten a monarch rarely attempt to eat another, or anything that looks similar!

Monarchs are among the best-known insect migra-tors. From northern breeding grounds, the adults fly south, where temperatures are milder and food more plentiful. Here they remain until winter passes, and then they return to their breeding grounds. Individuals rarely live long enough to make the entire migration; amaz-ingly, several different genera-tions negotiate a single migra-tion trek. On the California coast, monarchs often roost in Monterey pines and eucalyptus.

A. Monarch pupa
B. Monarch larva on milkweed
C. Painted lady
D. Monarch
E. Sphinx moth
F. Tiger swallowtail

Pacific Grove holds an annual festival to honor them.

Cabbage butterflies and **sulfur butterflies** (family Pieridae) are destructive to agriculture. Caterpillars of cabbage butterflies feed on plants in the mustard family, preferring cabbage and cauliflower but also eating wild mustard. The yellowish sulfurs sometimes occur in the thousands around large fields of alfalfa. The larvae feed on alfalfa, clover, and other plants in the pea family.

Swallowtails (family Papilionidae) are among the largest of our butterflies. The **western tiger swallowtail** (*Papilio rutulus*) has black stripes on a yellow background. The similar **pale swallowtail** (*Papilio eurymedon*) has a white background on its wings. Swallowtail caterpillars have eyespots and feed on willow, alder, and sycamore leaves. Swallowtails and other butterflies are speechless, but this does not stop them from using brilliant wings to announce themselves to their prospective mates.

WASPS, BEES, AND ANTS (ORDER HYMENOPTERA)

This group includes insects that sting and many of the social insects that form hives and colonies. They usually have two pairs of wings, although some are wingless, and most have chewing mouthparts. In this order, the female's ovipositor—the part that allows insects to insert their eggs into rocks, soil, and other protected places—may be modified into a stinger that injects venom.

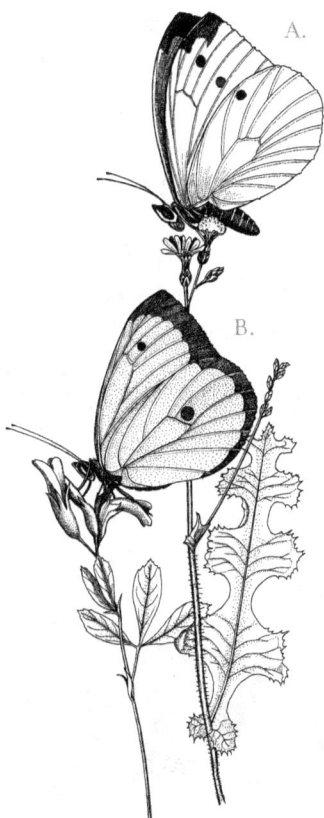

A.   Sulfur butterfly
B.   Cabbage butterfly

The tiny **gall wasps** (family Cynipidae) are seldom seen, but their presence is often evident because of the plant galls that they cause; a gnat-sized wasp can cause a gall the size of a baseball.

Despite their name, **velvet ants** (family Mutillidae) are actually wasps. The female is wingless and looks like a hairy ant. She deposits her eggs in nests of solitary bees, where larvae hatch and devour the larvae of the host. Her potent sting seems to have been devel-

Velvet ant

Oak gall formed by cynipid wasps

oped to subdue adult wasps and bees, but it has earned her such nicknames as cowkiller and mulekiller.

Yellow jackets (family Vespidae) are black-and-yellow, aggressive wasps that can sting multiple times and can induce shock in some people. When a worker is harmed, it releases an alarm chemical that alerts its fellow workers, which swarm to the defense of the nest. Their grayish-brown nests, recognizable to many people, are made of chewed wood mixed with saliva. **Paper wasps** also build nests, but these wasps are far less aggressive than yellow jackets.

Mud daubers belong to the family of spider hawks, or thread-waisted wasps (family Sphecidae), named for the obvious and often pronounced constriction between thorax

Yellow jacket

Paper wasp

and abdomen. The yellow-and-black mud dauber builds two to six mud cells covered by a sheet of mud, one of many nesting patterns in this family of wasps.

A female spider hawk builds the cells and lays an egg in each before sealing it. She hunts for spiders and paralyzes them with her sting so she can feed them to her larvae. Sphecid wasps will drag huge black widows

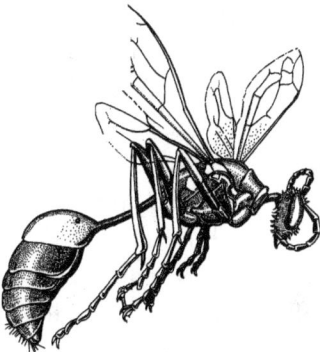

Mud dauber

along the ground if they are too heavy to carry. The large **tarantula hawk** (family Pompilidae) has a similar lifestyle but attacks tarantulas and other large prey.

A single fertile female **bumblebee** (family Apidae) starts an entire bumblebee colony. After overwintering in a protected spot, she emerges in the spring and finds an abandoned

Bumblebee

burrow, where she lays eggs. She collects pollen to feed these offspring, which develop into small females, called workers. The workers then provide food for the colony. In the fall, males, or drones, and fertile females, or queens, are produced. They then mate, and the cycle begins again.

Honeybee workers (family Apidae) are infertile females that collect nectar and pollen for food and defend the colony with their stings. Workers communicate the locations of rich food sources to their sisters by means of a waggle dance. The large queen lays eggs and is fed by the workers. When a hive becomes too large, a new queen is produced by special feeding, and the old queen leads a swarm of workers to establish a new hive. A swarm of bees is a fascinating sight. When they land, the queen is usually at the center. When beekeepers pick up swarms of bees, they must be certain to get the queen or the new hive will not survive.

California's docile strain of honeybee has hybridized with an aggressive Africanized strain that was accidentally introduced to Brazil in 1956 and has since moved northward. The **Africanized honey bee** is a subspecies of the **European**

Carpenter bee

honeybee (*Apis mellifera*) that tends to behave more aggressively than honeybees, swarming and stinging with little provocation.

Carpenter bees (subfamily Xylocopinae) are often seen buzzing around homes, searching for nest sites or pollen to eat. These hairy bees are robust and usually shiny black, though the males of some species are blond. They produce a lot of sawdust when excavating chambers. They are not social, although several may share a common entrance to their individual nests.

Leafcutter bees (family Megachilidae) differ from others by carrying collected pollen under their abdomens rather than on their legs. They

Africanized honey bee

Leafcutter bee

are called leafcutters because they slice small discs from leaves and use them to line the cells of their nests. One species, the **alfalfa leafcutter bee** (*Megachile rotundata*), is cultured for pollination of alfalfa, a crop that the shorter-tongued honeybee is not well suited to pollinate. Beekeepers provide wooden trays with holes for nest sites that the bees line with leaf discs.

Most **ants** (family Formicidae) are social and form colonies. They use chemical signals to recognize each other

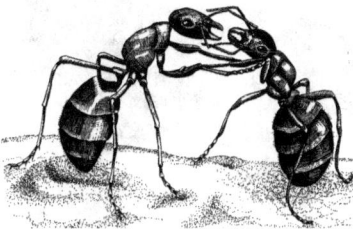
Ants passing food

and to form and follow trails to food sources. At somewhat unpredictable times but often after spring rains, the winged ants in a colony fly high in a swarm, where the males fertilize the females. After this mating flight the males die and the fertilized females disperse to begin new colonies. After digging a hole, each new queen bites off her wings and begins

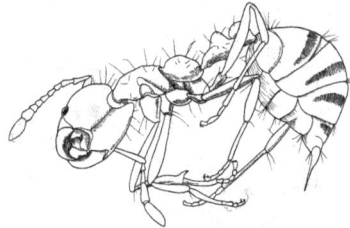
Fire ant

laying eggs. The eggs develop into workers that will feed the queen and future generations of young ants.

Most ants of the Valley/Foothill region are capable of biting, but the **red imported fire ant** (*Solenopsis invicta*) has the stinging ability found in primitive ants. They swarm and sting, causing intense pain and even death from anaphylactic shock.

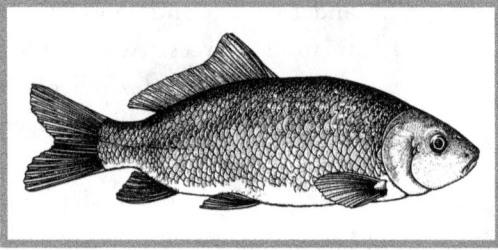

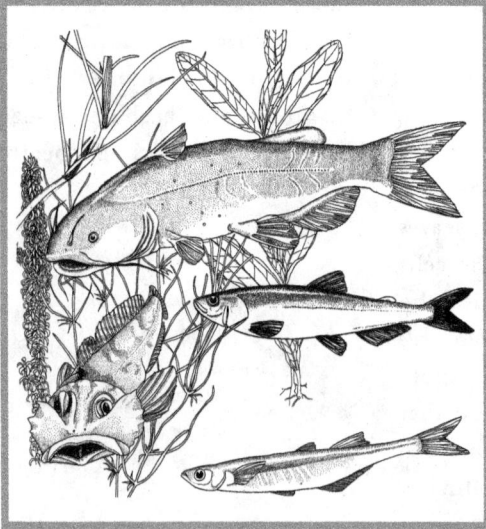

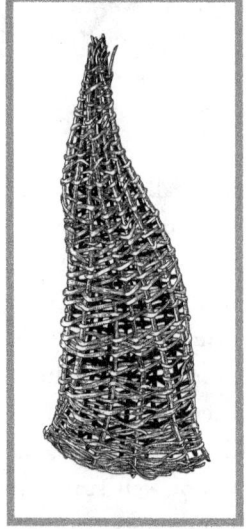

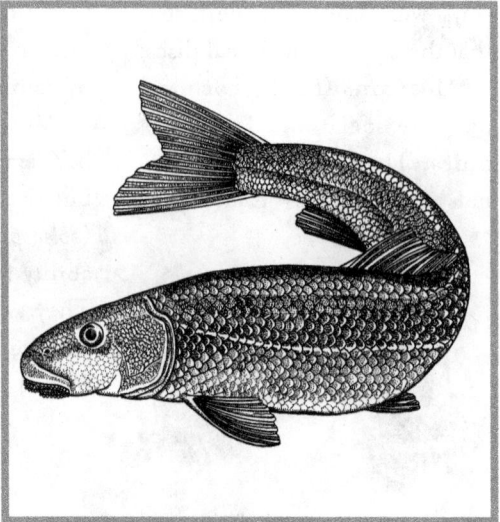

FISHES

Lampreys | Sturgeon | Shad | Salmon | Trout | True
Minnows (Pikeminnow, Hardhead, Golden Shiner,
Hitch, Roach, Carp) | Smelts and Silversides |
Mosquitofish | Stickleback | Sacramento Sucker |
Catfish | Striped Bass | Sunfish (Perch, Largemouth
Bass, Black Crappie, Bluegill) | Tule Perch | Riffle Sculpin

Fishes represent more species and more individuals than any other group of backboned animals. This is hardly surprising, considering that fishes inhabit water, a medium that covers over two-thirds of the earth's surface.

The manner in which a fish makes its living is revealed by the form of its body, including shape, mouth placement, fin placement, and other features. However, a fish is an adaptable creature and, when it comes to feeding and survival, will follow opportunity rather than a book it has not read.

Most fishes have streamlined bodies with two sets of paired fins, the pectoral and pelvic fins, that roughly correspond to our arms and legs. These paired fins help orient the fish and provide maneuverability. The other fins are unpaired and located on the midline. The fin on the tail, known as the caudal fin, provides forward thrust. The dorsal fin, along the back, and the anal fin, located under the belly, orient and stabilize the fish. The flexible rays and stiff spines of the fins support the fish and also help defend it.

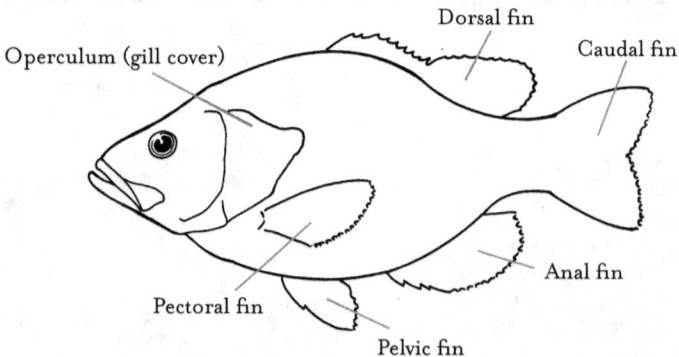

The swim bladder, a bag filled with gases, helps maintain buoyancy. A fish uses this organ to rise, sink, or float in place without a twitch from its fins. Fishes breathe with gills, usually four pairs. Bony fishes typically have a single gill opening covered by a bony plate, called the operculum. By alternately closing and opening its mouth and operculum, a fish pumps water—with oxygen—over its gills.

The bodies of most fishes are covered with scales of one of two types, called cycloid and ctenoid. Cycloid scales are usually round and thin with small rings. Ctenoid scales are similar but have many small, toothlike projections on the outer edge. It is not the size of a fish that reveals its age but rather the number of bands of rings on its scales. The layer of slime that fishes secrete from their skin cells helps fight fungal and bacterial infections and reduces drag by smoothing the fish's outer layer.

Many fishes rely on vision for finding food and mates, and for avoiding predators. Most freshwater fishes have color vision. A fish's eyes are capable of focusing at their very edges, allowing it to better detect motion. Many fishes have an excellent sense of smell. Fish nostrils are dead-end openings lined with sensory tissue; they are often divided by a delicate flap that separates inflow from outflow. Taste and smell are closely allied senses, especially in an aquatic environment. Some fishes, such as sturgeons and catfish, have whiskerlike skin extensions called barbels. Barbels have taste buds on them, and they also serve as organs of touch. On each side of a fish there is a row of sensory cells, called the lateral line, that senses low frequency vibrations and motion in the surrounding water.

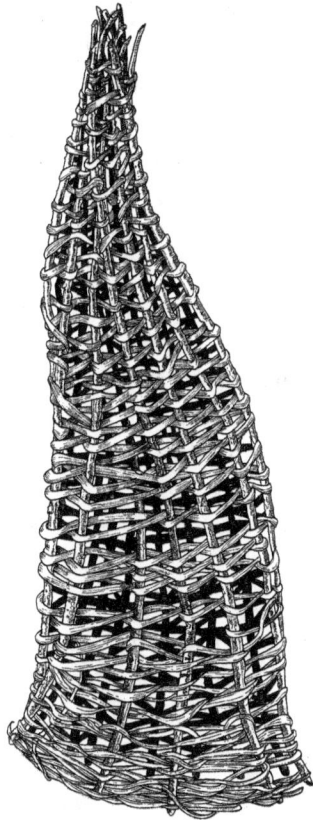

Fishing has long served as a window into the world of fishes. This fishing trap was woven from willow and wild rose by Miwok people.

Lampreys
(family Petromyzontidae)

Lampreys look like eels but are not closely related to them. They belong to the group known as the jawless fishes, or Agnatha. A lamprey has a disc-shaped mouth armed with formidable teeth, a skeleton of cartilage, seven paired gill openings, and a single nostril that opens on top of its head. Young lampreys have small eyes and round mouths without adult teeth.

A lamprey spends most of its life (four to seven years) as a larva, buried in a muddy stream bottom with its mouth exposed so it can filter nutrients from the water.

After becoming adults, parasitic species typically swim downstream, where they seek soft-scaled host fishes like trout and salmon. Upon finding a host, the lamprey attaches with its suckerlike mouth and rasps away the host's scales with its toothed tongue, eventually sucking out body fluids. The lamprey detaches when it has completed its meal. The host may survive or succumb to energy loss and infection. The parasitic adults return to streams to spawn after one to two years. They make nests in gravel, deposit eggs and sperm, and then die.

White Sturgeon
(family Acipenseridae)

The **white sturgeon** (*Acipenser transmontanus*) has the distinction of being the largest freshwater fish in North America. It reaches lengths of over twenty feet and weighs up to thirteen hundred pounds. These grayish fishes have a distinctive appearance: their elongate bodies are covered with bony plates, their tails look like sharks' tails, their mouths function like vacuum cleaners, and their snouts are distinctively blunt and triangular. They use their four barbels

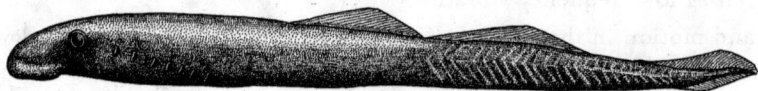

Pacific lamprey (*Entosphenus tridentatus*)

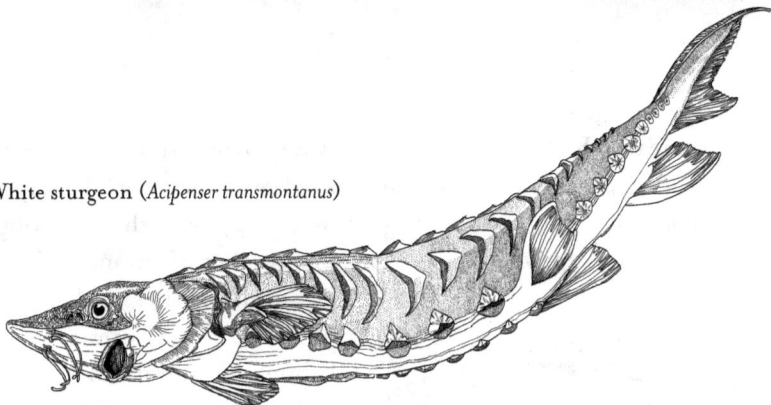

White sturgeon (*Acipenser transmontanus*)

to sweep the bottom, searching for prey such as invertebrates and fish.

White sturgeons live mostly in estuaries, with the only population in California occurring in the Sacramento–San Joaquin Delta. Sturgeons grow slow and mature late, spawning only after eleven years of age and sometimes not until twenty. Because their flesh and eggs are prized foods, they have been overharvested in the past and are still recovering.

Shad (family Clupeidae)

Native to the great Mississippi basin, **threadfin shad** (*Dorosoma petenense*) are small, schooling fish named for a threadlike projection at the base of the dorsal fin. They were introduced here as forage fish for game fish because they were

thought to feed in empty, open waters. Later studies found that they compete with young game fish for plankton. By the time their negative impact on game fish populations was understood, shad were firmly established here.

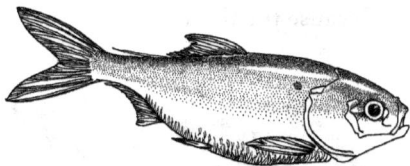

Threadfin shad (*Dorosoma petenense*)

Of the more than eighty species of fishes found in Valley/Foothill waters, at least thirty-seven are introduced. The effect of exotic species on fishery quality ranges from somewhat beneficial to extremely harmful, with the majority falling into the second category. Native fishes have declined drastically.

American shad (*Alosa sapidissima*) lives part of its life cycle in marine waters, reentering fresh water to spawn.

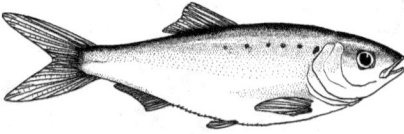

American shad (*Alosa sapidissima*)

Chinook Salmon
(family Salmonidae)

The **chinook** or king salmon (*Oncorhynchus tshawytscha*) is the largest California salmon. Because the flesh of this species is a gourmet treat, it supports a large commercial fishery as well as sport fishing, though the number of chinook salmon has declined due to damming of rivers, fishing pressure, and habitat deterioration.

Salmon are anadromous, with most chinooks now spawning in the fall, though subspecies have different spawning times. The drive to move upstream to spawn is impressive. Adults use visual and olfactory clues to return to the stream in which they were spawned. Spawning beds are in streams with cool, clean water and gravel bottoms.

Rainbow Trout
(family Salmonidae)

The **rainbow trout** (*Oncorhynchus mykiss*) is one of the most important game fish in California. Its adaptability, tasty flesh, and fighting spirit make it a favorite with anglers. Hatcheries throughout the state raise and stock this trout. Unfortunately, many native trout strains have been obliterated by interbreeding or competition with hatchery strains.

Chinook salmon (*Oncorhynchus tshawytscha*)

Steelhead are a strain of rainbow trout that migrate to the ocean to feed.

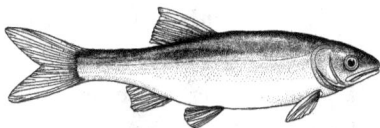

Sacramento pikeminnow
(*Ptychocheilus grandis*)

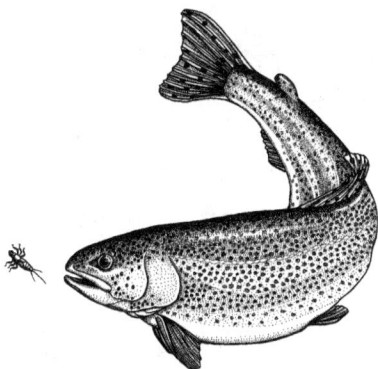

Rainbow trout (*Oncorhynchus mykiss*)

True Minnows
(family Cyprinidae)

Local members of the minnow family spend a majority of their lives near shore, feeding on plankton and small animals. Minnows lack teeth along their jaws: to determine whether or not a fish is true minnow, run your finger across the inside of its mouth.

The **Sacramento pikeminnow** (*Ptychocheilus grandis*) is a long predatory fish found in the clear streams of the foothills. For the most part, pikeminnows do not tolerate disturbance to waterways, nor to the composition of fish species inhabiting them. Before the introduction of many exotic fishes, pikeminnows were the top predators in many of the state's lowland waterways; pharyngeal teeth (knife-like projections in the throat) allow pikeminnows to eat prey including crayfish, insects, fish, birds, and rodents. They live in deep, shaded pools, where they tend to be sedentary, slowly finning over the bottom in search of prey. Pikeminnows frequently intermingle with the similar but smaller and less predaceous **hardhead** (*Mylopharodon conocephalus*).

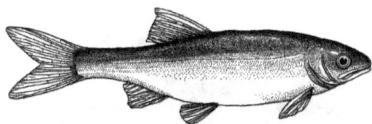

Hardhead (*Mylopharodon conocephalus*)

Many species in the minnow family originally entered California waters as live fish bait.

The **golden shiner** (*Notemigonus chrysoleucas*) and similar **hitch** (*Lavinia exilicauda*) often occur in slow-moving rivers, sloughs, reservoirs, and ponds. Their deep, compressed bodies, large scales, small heads, upturned mouths, and curved lateral lines all aid in identification.

Golden shiner
(*Notemigonus chrysoleucas*)

The **California roach** (*Hesperoleucus symmetricus*) is a minnow species that grows to five inches in length and has anal fin counts of eight to nine. The roach prefers cool, clear streams, unlike many other minnows that can survive in warm shallows.

Common carp (*Cyprinus carpio*), also in the minnow family, were introduced to California from Asia. They have large, yellow-brown scales and soft, fleshy lips with barbels. "Wild" **goldfish** (*Carassius auratus*) look like carp but without barbels. Carps' tolerance of warm, muddy water with low oxygen levels has allowed them to populate most of California's lowland waterways. They are considered by many to be trash fish and are often killed and discarded when caught. Carp meat has a muddy taste and contains many small bones, though with proper preparation it is valued by some as a delicacy.

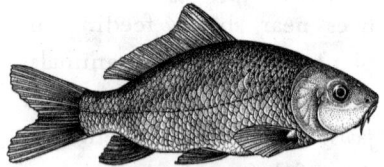

Common carp (*Cyprinus carpio*)

California roach (*Hesperoleucus symmetricus*)
Hitch (*Lavinia exilicauda*)

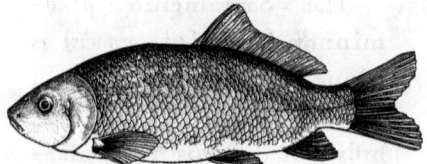

Goldfish (*Carassius auratus*)

Carp feed by shoveling their snouts into the mud, stirring up insect larvae and uprooting aquatic plants of shallow areas. This spoils such areas for use as nurseries by fishes that are considered more desirable. Carp are often blamed for spoiling waterways that are in fact degraded by human activity, when all they are guilty of is the ability to live in degraded habitats.

Smelts
(family Osmeridae)
and Silversides
(family Atherinopsidae)

There are many small, silvery fish that are not true minnows, even though they look like they should be. **Smelts** have an adipose (fatty) fin behind the main dorsal fin, a feature lacking in minnows and silversides. Several species thrive in brackish

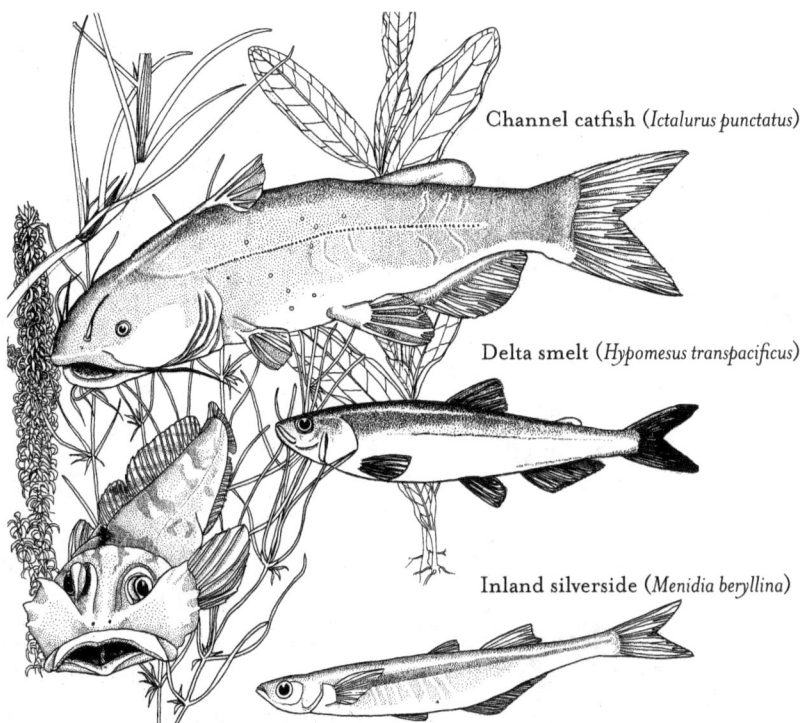

Channel catfish (*Ictalurus punctatus*)

Delta smelt (*Hypomesus transpacificus*)

Inland silverside (*Menidia beryllina*)

Riffle sculpin (*Cottus gulosus*)

waters of the Sacramento–San Joaquin Delta, where they are prey for striped bass and other cruising predators. Long, silvery fish in the Delta that lack both teeth and adipose fins are likely to be silversides. The introduction of **inland silversides** (*Menidia beryllina*) into inland waters has disrupted habitats of native fishes; initially introduced to control the pesky midge fly by eating its larvae, silversides now eat food that once supported native and game fish.

Mosquitofish
(family Poeciliidae)

Like their relatives the guppies, **mosquitofish** (*Gambusia affinis*) are livebearers; eggs remain in the female's body until they hatch. This strategy limits the number of eggs produced in each brood, but it also allows the female's body to protect

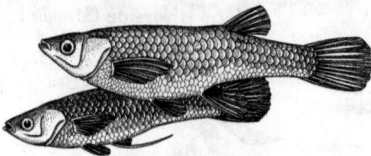

Mosquitofish (*Gambusia affinis*)

and incubate the eggs. Tail fins of both sexes are rounded, and coloration is uniformly gray with an iridescent sheen. Mouths open upward, adapted for surface feeding. These fish sometimes gulp air to survive in waters with low oxygen levels.

Mosquitofish were introduced from the Midwest to California to control mosquitoes. They do eat many mosquito larvae and pupae but are omnivorous, capable of subsisting on a variety of foods. Therefore they are an excellent tool for mosquito control, but they compete with native fishes. Their large numbers and shoreline habitat make western mosquitofish the most often seen California fish.

Stickleback
(family Gasterosteidae)

Sticklebacks are small fish armed with a series of stiff spines and, in place of scales, a number of bony plates. The male **threespine stickleback** (*Gasterosteus aculeatus*) establishes his territory and makes a nest by collecting and "gluing" veg-

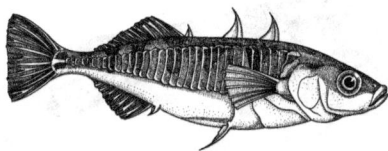

Threespine stickleback
(*Gasterosteus aculeatus*)

etation into a stack through which he wiggles, creating a tunnel. He then, in breeding coloration, does a zigzag display for passing females. If she is willing to breed, a female enters the nest tunnel and lays eggs, which the male fertilizes. The male also fans the eggs and guards the young. After he chases away the first female, he may court several more.

Sacramento Sucker
(family Catostomidae)

Sacramento suckers (*Catostomus occidentalis*) are specialized bottom feeders. With fused, fleshy lips on the undersides of their heads, they "vacuum" the bottom. When they are young, these fish eat mostly invertebrates. As they mature, they eat more and more algae, reaching an adult ratio of approximately 80 percent algae to 20 percent invertebrates.

Decaying ooze also contributes to the diet of suckers. Like squawfish, they have pharyngeal teeth (bony projections in the throat) that prevent large particles from entering the digestive tract. Their flesh, though somewhat bland and bony, was traditionally preferred by some native people over trout.

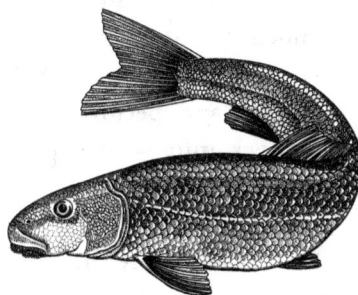

Sacramento sucker
(*Catostomus occidentalis*)

Catfish
(family Ictaluridae)

Catfish can be recognized by their eight sensory barbels, their large heads, and their scaleless skin. They also have an adipose fin along the back. All of California's catfish have been introduced from the eastern United States. In general, catfish prefer to feed away from light, often foraging after

Black bullhead catfish
(*Ameiurus melas*)

sunset. They tolerate low oxygen levels, often by gulping air at the surface. Catfish are well defended from predators by three stout spines, one each in the dorsal and pectoral fins, which lock into position and are covered by skin containing poison glands.

Catfish are unusually good parents, as California fishes go. After the nest is made and the eggs are laid and fertilized, both parents guard and fan the eggs. When the eggs hatch, the parents continue to guard the tiny fry until after the yolk sac is absorbed, sometimes for two weeks after hatching. Parental care is a factor in the success of catfish.

introduced from the East Coast in 1879. A striped bass requires a large river in which to spawn and incubate its eggs, an estuary with invertebrates upon which the young can feed, and plentiful forage fish in a large body of water. Unlike the many other anadromous fishes that die after a single spawning, the striped bass can make the challenging adjustment between fresh and salt water multiple times.

Striped bass are group spawners, forming aggregations in April from which smaller groups of mostly males and a few females separate and spawn. Fertilized eggs drift in

Striped Bass
(family Moronidae)

The **striped bass** (*Morone saxatilis*) is a robust fighting species

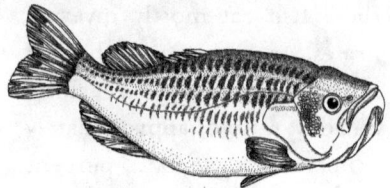

Striped bass (*Morone saxatilis*)

the current, dying if they lodge on the bottom and otherwise hatching in about two days. Striped bass, like salmon, have declined alarmingly in recent years, due to water diversion and reduced habitat quality.

Sunfish
(family Centrarchidae)

The **Sacramento perch** (*Archoplites interruptus*) is the only native member of the sunfish family in the West. Originally they occurred only in California, in the Sacramento–San Joaquin river system. They evolved long ago and survived to the present in part because of their ability to tolerate alkaline waters. In their native waters,

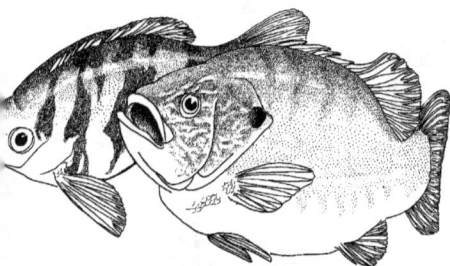

Sacramento perch (*Archoplites interruptus*)
and green sunfish (*Lepomis cyanellus*)

Sacramento perch have largely been displaced by introduced members of the sunfish family.

Largemouth bass (*Micropterus salmoides*) are often the top predator in the warm waters they occupy. Older bass often eat fish, but largemouths sometimes prefer a specific food type, sometimes a particular fish species, sometimes crayfish, tadpoles, or frogs.

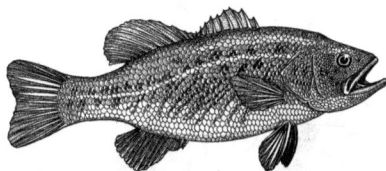

Largemouth bass
(*Micropterus salmoides*)

Adult largemouth bass are solitary fish that typically occupy a territory where they hunt from ambush. They also may rove when hunting. The initial hit of a largemouth feeding on surface prey is often like a small explosion. Two other bass, the **smallmouth bass** (*Micropterus dolomieu*) and the **spotted bass** (*Micropterus punctulatus*), favor rivers and reservoirs, respectively.

Two other prominent introduced sunfish are black crappie and bluegill. **Black crappie** (*Pomoxis nigromaculatus*) seek shelter around submerged vegetation. The young consume mostly plankton, while the adults supplement this with small fishes. Their checkered pattern led to their nickname, calico bass. Black crappie build nests and the males guard the eggs and protect the young for a short period after they hatch.

Black crappie
(*Pomoxis nigromaculatus*)

Bluegill (*Lepomis macrochirus*) are the most common warm-water game fish in California. Introduced from the eastern United States, bluegill are able to survive and reproduce in a variety of conditions.

Green sunfish (*Lepomis cyanellus*) are more streamlined than bluegill and have rounded pectoral fins. These two species are the fish most likely to nip at barefoot toes dangled off a boat dock. They feed on virtually any prey of manageable size and eat algae when nothing else is available. Both species make nests and defend territories in the breeding season.

Tule Perch
(family Embiotocidae)

Tule perch (*Hysterocarpus traskii*), a native fish, live in large, lowland waterways, including

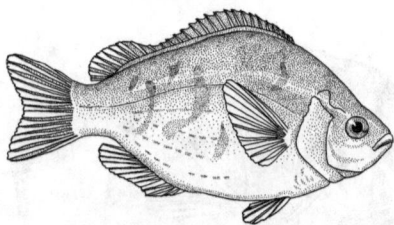

Tule perch (*Hysterocarpus traskii*)

tule beds. They can be identified by their distinctive fins, round body shape, and a row of enlarged scales at the base

Bluegill (*Lepomis macrochirus*)

of the dorsal fin. Both barred and unbarred forms of this fish exist. Their grinding teeth are adapted for feeding on hard-shelled, bottom-dwelling clams, snails, and insects.

Riffle Sculpin
(family Cottidae)

Riffle sculpins (*Cottus gulosus*) have a distinctive wide-headed, pop-eyed, mottled appearance. Their cryptic coloration and irregular outline make them nearly invisible on the rocky bottoms of streams where they dwell. Because they lack a buoyant swim bladder, they are superbly adapted to a life of feeding on bottom invertebrates.

Sculpins generally avoid light by feeding at night and taking refuge by day under rocks or in other protected places. Reproduction occurs in the spring, when the territorial male constructs a nest under a rock by excavating an opening. Females are escorted into this cavity, where the eggs are laid and fertilized. Young sculpins typically drift downstream and establish residency.

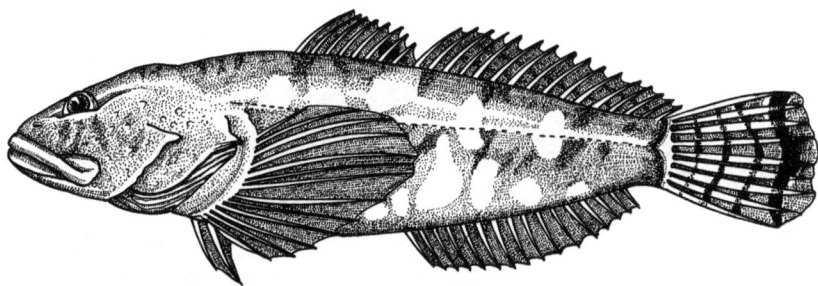

Riffle sculpin (*Cottus gulosus*)

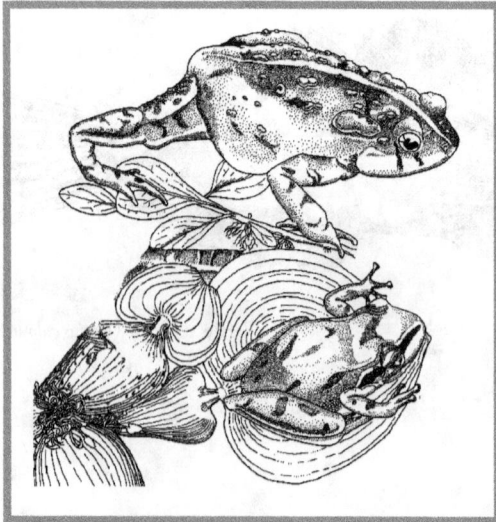

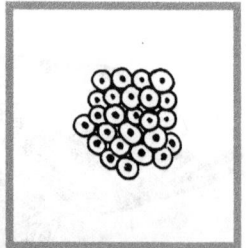

AMPHIBIANS

Salamanders and Newts | Frogs and Toads

For years, the first serious encounter many students had with an amphibian was with a frog in the dissecting tray in high school biology. The fact that the frog was lifeless, faded, and embalmed in smelly preservative created an unpleasant impression of amphibians.

The reality of these marvelous animals is very different and takes many forms, including the spring cacophony of courting frogs, the excitement of a child at the capture of her first frog, a pond teeming with tadpoles, or a flooded pasture alive with hopping frogs. Because there are so few amphibians in the Valley/Foothill region (less than a dozen species), this is an easy group to learn.

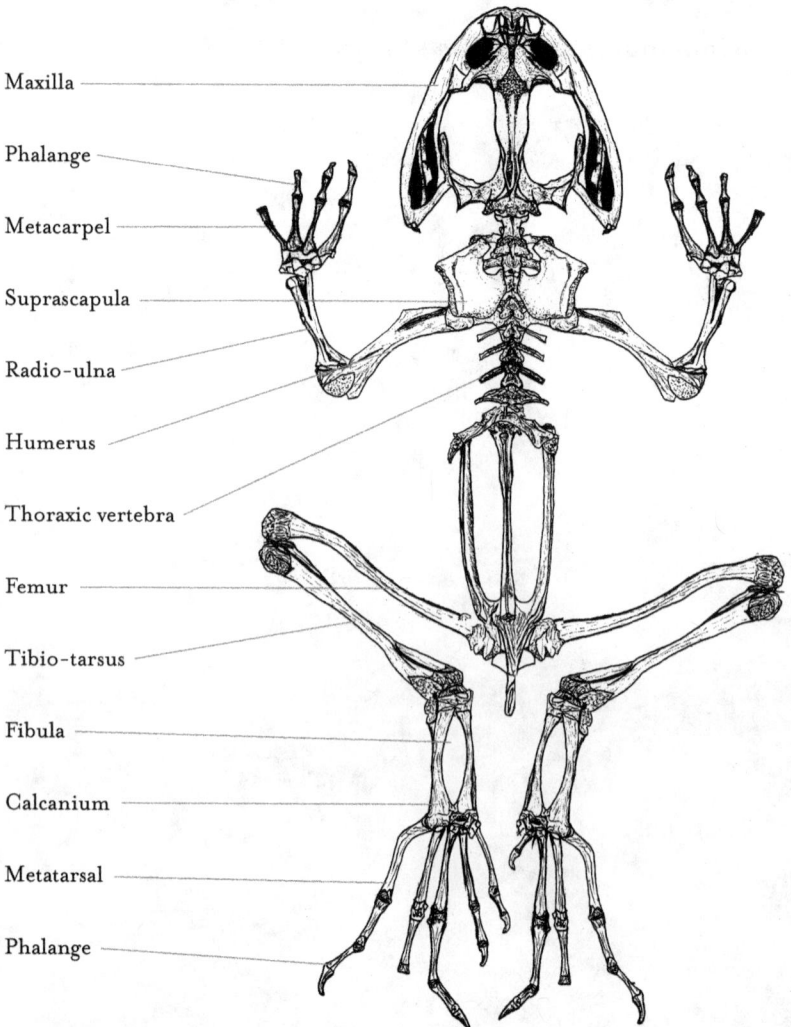

Maxilla

Phalange

Metacarpel

Suprascapula

Radio-ulna

Humerus

Thoraxic vertebra

Femur

Tibio-tarsus

Fibula

Calcanium

Metatarsal

Phalange

Amphibians are backboned animals with moist, glandular skin. Thin and smooth, this skin makes them vulnerable to drying out. Some toads have thickened skin with a rough, bumpy surface that is caused in part by natural waterproofing materials. Such skin allows greater independence from water but does not eliminate the need to avoid hot, dry environments. The semipermeable skin that allows water to escape also allows oxygen to enter, so an amphibian can literally breathe through its moist skin. This is important because many amphibians have lungs too small to provide enough oxygen for their body size. Amphibians are ectothermic, meaning that their body temperature and metabolic rate are dependent on the temperature of their environment.

The roots of the word *amphibian* translate to "double life," and indeed the change from a water-dwelling, gilled larva to a land-dwelling, air-breathing adult does involve a double life. Because amphibian eggs are gelatinous and lack a shell, most must be deposited in water to allow development. Each species does this in a distinctive way. Some eggs are in long, floating strings, others are in clumps, and still others are deposited individually. Some amphibian species are able to lay their eggs on land in protected, humid locations.

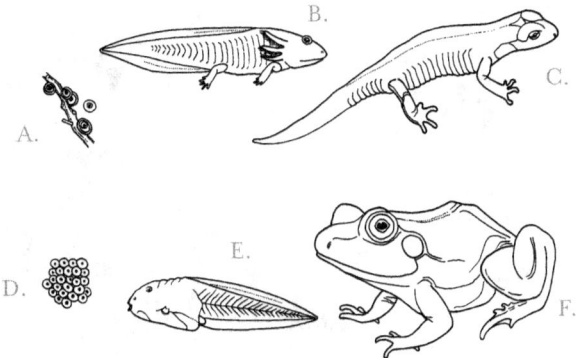

AMPHIBIAN DEVELOPMENT

A. Eggs (on a stick)
B. Larva
C. Adult salamander

D. Egg mass
E. Larva (tadpole)
F. Adult frog

The majority of amphibians are found near rivers, marshes, and irrigation ditches. But because several local amphibians are adapted to a terrestrial lifestyle, surprisingly small bodies of water can support amphibian breeding. Most of these animals, notably western spadefoots and California tiger salamanders, depend on rain-filled depressions—vernal pools—for their breeding habitat.

California tiger salamander (*Ambystoma californiense*) near a vernal pool

Salamanders
(order Caudata)

Salamander is a general term for any amphibian with four legs and a tail at maturity. During reproduction, the male produces a packet of sperm, the spermatophore, which the female picks up and deposits in her own body, often following a courtship dance by the male. The sperm may be used to fertilize eggs immediately or they may be stored for future use.

The larvae of salamanders are elongate, have external, filamentous gills, and develop legs very quickly. Unlike frog larvae, salamanders develop front limbs first. Larval salamanders are carnivorous.

The **California tiger salamander** (*Ambystoma californiense*) has pale yellow spots on an inky-black body. Tiger salamanders are seldom seen, in spite of their conspicuous size and coloration, because they have secretive habits. They spend much of their time underground in rodent burrows, under rocks, or in cracks, earning, with the rest of their family (Ambystomatidae), the common name of "mole sal-amanders." Adults are seen during rainstorms in late January as they travel to vernal pools and streams to breed. The larvae feed on snails and tadpoles. Adults eat earthworms, small invertebrates, amphibians, and even mice.

The **California newt** (*Taricha torosa*) is most evident during its breeding season in foothill streams. Its bright orange belly contrasts with its darker, rust-colored back. The skins of many salamander species contain toxins that discourage predators. Some are among the most potent naturally occurring toxic substances known. Such salamanders are often brightly colored and boldly patterned, perhaps to advertise their distastefulness to predators. The California newt, whose skin toxins are extremely potent, arches its back to display its colorful underside as a warning when threatened. Adults live on land most of the year, returning to water to breed late each winter. California newts are rough-skinned during the terrestrial stage, becoming smoother as they return to breeding waters.

Smaller in diameter than a

pencil, the **California slender salamander** (*Batrachoseps attenuatus*) is sometimes called a worm salamander because of its small girth, tiny limbs, and costal grooves, which provide an almost segmented appearance. The California slender salamander is one of several species that live mostly under rocks, logs, and damp litter in Foothill woodlands. This little salamander is completely terrestrial, even breeding and laying eggs on land—an amazing feat considering that the slender salamander does not have lungs, but breathes entirely through its moist skin.

Ensatinas (*Ensatina eschscholtzii*) are most likely to be encountered during the first rainy weeks of fall. At this time, the young gather in

Ensatina (*Ensatina eschscholtzii*)

rain-filled pools to feed on worms and other small animals in and around water. The adults become more secretive, remaining mostly under logs and in other shaded habitats where ensatinas hunt, mate, and carry on most of their adult activities.

Frogs (order Anura)

These are the most familiar amphibians, conspicuous for their voices that produce a lovely serenade each spring. Such calls are important for attracting mates and establishing territories.

The order name, Anura, means "having no tail." Frogs have tympana (ears) and true voice boxes. Males, which do most of the calling, usually have one or more inflatable vocal sacs on their throats that serve as amplifiers. Fertilization is typically external and accomplished in the water while the male is grasping the female around the waist or chest. The male releases sperm as the female lays eggs.

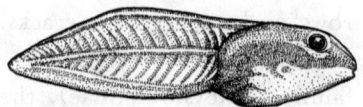

Tadpole

Frog larvae are called pollywogs or tadpoles, a word derived from roots meaning "toad head." Metamorphosis is drastic in frogs because they change their mode of breathing and their body shape and they switch from a mostly vegetarian to a carnivorous diet. In the first several days after hatching, a covering develops to enclose a tadpole's gills. Tadpoles of Valley/Foothill frogs are mostly herbivorous, beginning life by scraping algae and scum from the bottom of the pond.

Scientists have noted a decrease, possibly worldwide, in frog populations, the cause of which is currently not known.

Toads, by virtue of their thicker, bumpy skin, are more independent of water than most frogs, but they still must breed in water. Their call is a mellow chirping. Females may lay as many as ten thousand eggs per year! Many of the resulting

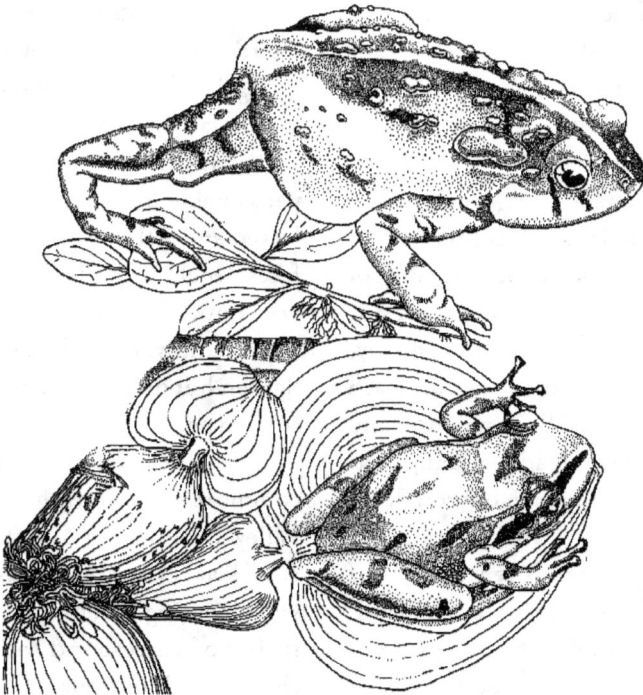

Western toad (*Anaxyrus boreas halophilus*)
Northern Pacific chorus frog (*Pseudacris regilla*)

toadlets serve as food for other animals.

Toads often hunt under porch lights and streetlights, where they establish a pecking order. Hunting under streetlights, they may also be flattened by cars.

Toads do not cause warts, but their skin does produce toxins that help protect them from enemies. A large poison gland is located behind each eye.

A common garden and wildland resident, the **western toad** (*Anaxyrus boreas halophilus*) is a stocky creature that often spends its days in an underground retreat. In the cool of evening, it emerges with a voracious appetite and forages for insects and similar small animals.

The **western spadefoot** (*Spea hammondii*) is named for a unique feature: a horny, wedge-shaped spade on each rear foot. Backing up and rotating its spade-bearing feet, the spadefoot can burrow down about twenty inches, where it typically spends ten months of each year. During this period, it may survive the loss of up to 60 percent of its bodily fluids. The loss of about 15 percent kills most mammals.

Spadefoots emerge in late winter to breed in vernal pools, where the calling males sound like they are snoring or sawing. Because of the temporary nature of these pools, western spadefoots develop quickly, hatching in one day and changing from tadpoles to adults in less than three weeks. Spadefoots differ from true toads by having vertical pupils in their eyes and smooth skin, and by lacking the toad's poison glands.

The slender **northern Pacific chorus frog** (*Pseudacris regilla*) has a dark eye stripe and enlarged pads on the tips of its fingers and toes. These pads enable the frog to cling to leaves and other smooth surfaces (including glass). There are both brown and green forms; individuals darken and lighten but do not change basic color type.

The chorus frog is the most commonly heard frog in the Valley/Foothill region. Its call is a "krik-ek" with the second syllable rising. Evenings in late spring often resound with the chorus of hundreds of males gathered at ponds or other bodies of water, where they

Western spadefoot (*Spea hammondii*) habitat

stake out territories with their creaking calls and compete to attract females.

Once the common frog of shaded pools and reedy marshes in the lowlands of California, the **northern red-legged frog** (*Rana aurora*) is now quite rare. They were hunted by the thousands during the gold rush of the past century and later their legs were marketed as food. However, it was not until the American bullfrog was introduced to California that a serious decline in red-legged frog populations occurred.

This medium-sized frog has unwebbed front feet, webbed hind feet, and red coloration on the underside of its body and hind legs. Its voice is a series of soft, accelerating, guttural notes often ending in a growl.

The **foothill yellow-legged frog** (*Rana boylii*) is patchily distributed in the Sierra foothills, occurring mostly near clear streams where the habitat is relatively natural and not overrun with bullfrogs. It is the only frog in the foothills with a yellow pattern on its belly and the undersides of its legs.

The largest frog in North

Northern red-legged frog (*Rana aurora*)
American bullfrog (*Rana catesbeiana*)

America, the **American bullfrog** (*Rana catesbeiana*) was introduced to California from the eastern United States. Bullfrogs are commonly encountered along Valley/Foothill waterways, where they reveal themselves with a birdlike alarm call sounding like "erk," followed almost immediately by the splash of the diving frog. The announcement call of the bullfrog is also frequently heard: a deep, resounding "jug-o-rum" that is easy to mistake for the vocalization of a much larger animal—like a bull. The male emits its deep, rumbling calls to attract females and to establish his territory. Each night in the warm months, the male returns to a specific calling station.

Smooth, green, mottled skin, large eardrums, and a fold of skin extending from the eye around the eardrum help identify the bullfrog. An adult bullfrog eats nearly any moving thing that will fit into its huge mouth, including insects, fish, mice, other frogs, birds, and snakes. These aggressive frogs are widely distributed and can be seen and heard throughout most of California, including such unlikely habitats as urban canals and desert pools.

Foothill yellow-legged frog
(*Rana boylii*)

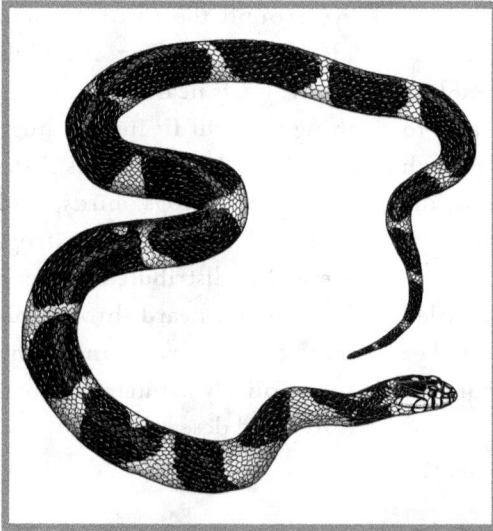

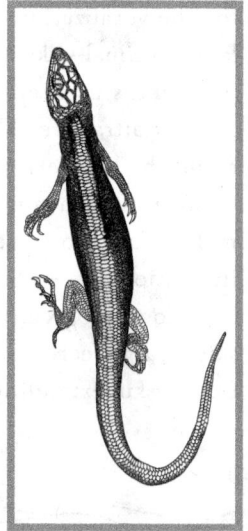

REPTILES

Turtles | Lizards | Snakes

Humans, although rarely fond of reptiles, are often fascinated by them. The prehistoric appearance, mythical power, and interesting life histories of these scaly creatures all attract us.

Reptile scales are part of the skin and, unlike fish scales, cannot be removed without tearing the skin. The scales of some reptiles, such as turtles and alligators, grow and are not shed. But some reptiles must shed their scaly skins periodically to accommodate their growing bodies, which is why entire cast skins are sometimes found tangled in rocks where snakes have rubbed them from their bodies.

It is this scaly exterior that allows most reptiles to live away from moist habitats, protecting them from heat and dehydration. Further independence from water is gained by reptiles' ability to lay eggs with shells. Unlike amphibians, young reptiles can develop in eggs deposited in protected places on land. (Some reptiles are "born alive" from eggs that are without shells and are incubated within the mother's body.) The membranes and shell that surround the developing young prevent drying out, facilitate breathing, and store waste. Food is provided in a yolk sac that shrinks as the young animal grows. Reptiles have become so terrestrial that some otherwise fully aquatic animals—sea turtles, for example—must return to land to lay eggs.

Reptiles are often thought of as cold and slimy; most are neither, if they can help it. Birds and mammals elevate their body temperatures through internal metabolic reactions, but reptiles must derive heat from their environment, a process called ectothermy. Some reptile species maintain an active temperature of over 100 degrees Fahrenheit—hardly cold! A lizard can maintain an almost constant temperature by moving between sun and shade, by changing the pattern of its blood flow, and by altering the posture and color of its body. Using these techniques, it can accumulate heat and actually achieve a body temperature warmer than its environment. This requires sunshine, and therein lies the major disadvantage of ectothermy; it is weather-dependent and therefore uncertain. When the weather is unfavorable, most reptiles are inactive.

Because their body temperature is independent of metabolism, reptiles are spared the necessity of constantly eating large amounts of food. This is a significant advantage, especially in habitats with low or irregular food supplies, such as deserts, where reptiles are often the dominant vertebrate group. Reptile diversity is greatest in the warm, stable tropics and decreases through the temperate zones toward the poles.

Some reptiles use venom (modified saliva) to capture prey or to protect themselves. Snake venoms were once classified as hemotoxic, neurotoxic, or cardiotoxic, depending on whether the venom primarily affected the blood, the nervous system, or the heart of the victim. It is true that proteins in venom produce a poisonous effect on a specific bodily organ or system, but we now know that one venom can affect several organs simultaneously.

Newly hatched venomous snakes possess potent venom, and although the volume may be less than in mature snakes, it is often enough to be lethal. Evidence indicates that snakes can inject half of their venom or less per strike, debunking the idea that snakes strike only once with potent venom. Rattlesnakes are the only dangerously venomous snakes occurring in California.

Turtles
(order Testudines)

The hallmark of this group is the shell, a structure formed by a layer of bone covered with large scales. Turtles cannot jump, fly, or run with this cumbersome armor. Also, the shoulder joint of a turtle is inside the turtle's ribs, a unique anatomical contortion required to accommodate the shell. Young turtles, while still inside the egg, have a "normally" located shoulder joint, but as the shell develops it forces the shoulders back inside the ribs. Turtles lack teeth; they use their tough beaks to consume an omnivorous diet.

Turtles have often been the subject of folk tales. Such tales depict turtles as being ancient and wise, as they indeed appear to be. In fact, some turtles are known to live to over one hundred years of age, with some poorly documented cases exceeding one hundred and fifty years. But the wisdom of turtles is a myth; they have small brains and are typically guided by instinct.

Western pond turtles (*Actinemys marmorata*) are the only Valley/Foothill native, although you may sometimes see released pets, such as the **red-eared slider** (*Trachemys scripta elegans*). Pond turtles live near deep pools. Often seen sunning themselves on logs or rocks, sometimes with mouths agape, these turtles escape into the water and swim to the bottom when threatened. They can stay underwater for extended periods by absorbing oxygen from the water through their mouths and anal linings.

Western pond turtles eat plants, small animals, and carrion. Mating occurs underwater. The female digs a hole on the bank or sometimes on a hillside far from water and lays her eggs there. Formerly quite common, this turtle is now a species of special concern due to

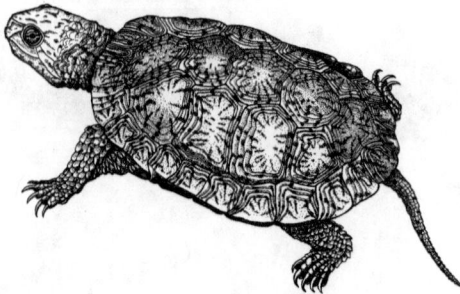

Western pond turtle
(*Actinemys marmorata*)

its dwindling numbers—reportedly down to less than 1 percent of their former population.

Lizards and Snakes
(order Squamata)

A lizard has four legs with claws, a tail, ear openings, and moveable eyelids—a list that distinguishes them from both salamanders and snakes. Even the **northern California legless lizard** (*Anniella pulchra*) exhibits basic saurian features, although it resembles a snake.

Some lizards have the unsettling ability to detach their own tails when threatened.

Western fence lizard
(*Sceloporus occidentalis*)

The tail then wiggles for several minutes, distracting the predator and allowing the lizard to escape. The break occurs along special fracture planes in the bones and muscles of the tail. Because lizards store fat in their tails, they live longer if they keep them, and they also lay larger eggs: be gentle when handling lizards.

Western fence lizards (*Sceloporus occidentalis*) are common occupants of woodpiles, trees, sheds, rock piles, and fence posts. Found in diverse habitats from sea level to high elevations, western fence lizards are variable in color and markings, but they often have blue patches on their bellies and are, in fact, commonly called "bluebellies." The dark scales on their backs are keeled, giving these lizards a rough texture.

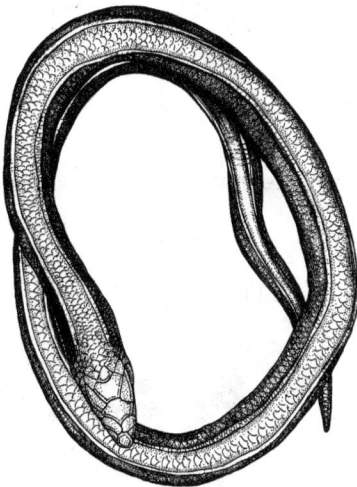

Northern California legless lizard
(*Anniella pulchra*)

The **common sagebrush lizard** (*Sceloporus graciosus*), and **desert spiny lizard** (*Sceloporus magister*) are related to the fence lizard and also have blue belly patches. Males often engage in ritualized displays, doing pushups to show their bright bellies, in order to attract females and intimidate rival males. Because a high perch is a sign of dominance among males, scrambling competition for such desired perches often results in amusing antics. These lizards feed on insects and spiders. If one loses its tail, the tail can keep wiggling for up to an hour.

Gilbert's skinks (*Plestiodon gilberti*) are extremely smooth lizards. Young skinks are striped and their tails are bright blue, or sometimes reddish. As these lizards mature, their stripes and tail coloration fade, although their tails and heads can turn a vivid orange-red in the spring breeding season. Adults are plain brown, sometimes with indistinct spots. The young hatch from eggs shortly after they are laid. Insects and spiders are an important part of the Gilbert's skink's diet.

Gilbert's skinks can be

Gilbert's skink (*Plestiodon gilberti*)

found in grassy Foothill areas by investigating rustles in spring and summer. To escape, a Gilbert's skink presses its limbs against its body and wriggles through grass in a manner similar to a snake.

Western skinks (*Plestiodon skiltonianus*), striped as adults, range along the coast and inland as far south as Lake Tahoe.

Western skink (*Plestiodon skiltonianus*)

Although the **tiger whiptail** (*Aspidoscelis tigris*), named for its long, tapered tail, is streamlined and fleet of foot, its walk is jerky, with a characteristic side-to-side movement of the bullet-shaped head. The whiptail's powerful hind legs propel it to great speeds in open habitats, enabling it to capture flies and other airborne insects. Whiptails are said to eat the eggs of birds and of other reptiles, possibly locating such food items by flicking their tongues to sample airborne chemicals.

Tiger whiptail (*Aspidoscelis tigris*)

The **southern alligator lizard** (*Elgaria multicarinata*) is the largest and most imposing of the Valley/Foothill lizards. Large scales, a long tail, a blocky head, and a willingness to swim gained it the "alligator"

in its name, although in the lizard's case the urge to swim involves escaping from predators. Crossbars in its pattern and pale yellow eyes are other distinguishing features. The alligator lizard will aggressively defend itself, biting and twisting ferociously while smearing feces, but its bite does us no real harm.

Southern alligator lizards eat insects, spiders, lizards, and small mice. Like other lizards of this region, they hunt moving prey. Before striking, the alligator lizard moves its head from side to side to estimate the distance. Usually active in the day, the alligator lizard may become nocturnal later in the summer as temperatures rise.

Southern alligator lizard
(*Elgaria multicarinata*)

Common side-blotched lizards (*Uta stansburiana*) are abundant in dry, sandy places throughout the West. They take refuge in animal burrows or other holes, reaching safety with a quick dash. A few minutes of sunshine will entice them back out to bask on the surface in all but the coldest months of the year. The distinctive inky-blue blotch behind each foreleg identifies this lizard, although it is sometimes faded or absent. Side-blotched lizards eat insects, spiders, pill bugs, and similar food.

Common side-blotched lizard
(*Uta stansburiana*)

Often in the headlines of Central Valley newspapers because of its endangered status, the **blunt-nosed leopard lizard** (*Gambelia sila*) is a large lizard with a good-sized head and a round tail. Its body is yellowish or grayish with many dark spots. This lizard is endemic to central California, living on the west side of the southern San Joaquin Valley, an area where big agriculture and land development are causing rapid change.

Leopard lizards are active only when surface temperatures get quite warm (near 80°F), otherwise remaining inactive in burrows. They are aggressive predators, feeding on insects—especially grasshoppers—and smaller lizards, and sometime even eating plants. When threatened, the leopard lizard may run upright on its two large hind legs and seek refuge in a mammal burrow.

Blunt-nosed leopard lizard
(*Gambelia sila*)

The **coast horned lizard** (*Phrynosoma blainvilli*), or horny toad, has a distinctive appearance but is so well camouflaged that it is hard to see! Its flat, squat body lies so low to the ground that it does not produce a revealing shadow. This built-in protection is enhanced by the lizard's habit of partially burying itself in loose sand when the sun sets. Its nostrils are equipped with valves that can close in order to keep sand out. The impressive array of horns and spiked scales presents a formidable defense to predators, and it breaks up the lizard's outline as well. An inactive hunting style also helps to keep this lizard out of sight. Horned lizards are often found near anthills, eating voraciously. Even new hatchlings about the size of a dime can be found eagerly gulping down ants.

When threatened, the horned lizard inflates its body with air, making itself difficult to swallow. If that fails to discourage the enemy, it squirts blood from its eyes through pores in the eyelids! This once common lizard is now quite rare in the Valley/Foothill region due to habitat loss.

Coast horned lizard
(*Phrynosoma blainvilli*)

The snake, perhaps because of its unblinking stare and the existence of venomous species, is the reptile most feared by humans. Thanks to a double-jointed hinge in the jaw and a flexible muscle joining the halves of the lower jaw, snakes have the alarming ability to swallow prey of larger diameter than themselves. Sometimes a snake's ambition exceeds its abilities, as in the account of a two-and-a-half-foot garter snake that attempted to swallow an adult bullfrog. The snake partially engulfed one of the frog's legs and was thoroughly thrashed about before escaping to fight another day.

Some snakes possess an organ in the roof of the mouth that detects chemicals—and therefore food—in the environment.

This organ is the reason they flick their tongues. All snakes are carnivorous, and many are significant natural controls on prey species. Snakes are often killed by people who are driven by fear, unaware of the snake's importance to the balance of nature. They are also often killed inadvertently: on cool mornings or after sundown, snakes often stop on warm road surfaces—an excellent way to speed up the metabolism for hunting, but a hazard that the snake cannot comprehend.

The slender **common garter snake** (*Thamnophis sirtalis*) supposedly resembles an old-fashioned garter, having a central yellow stripe and a yellow stripe on each side. These snakes are generally found along the banks of ponds and rivers. Amphibians and their larvae compose the majority of this snake's diet, supplemented with earthworms, fish, mice, and insects. If caught by a predator, a garter snake will often release a smelly discharge from its vent and try to smear it on its captor. Reproduction occurs by live birth, with the female typically bearing twelve to eighteen offspring; broods of up to eighty-five have been reported. The **giant garter snake** (*Thamnophis gigas*) is a relative that is threatened due to the loss of inland marsh habitat.

Common garter snake (*Thamnophis sirtalis*)

The **California striped whipsnake** (*Coluber lateralis*) has a yellow or sometimes orange stripe extending along each side. One interesting trait of the whipsnake is its tendency to elevate its head high above the ground while hunting. Whipsnakes are associated with Foothill habitats, sometimes far from water.

Called the king of snakes because of its dominion over other snakes, the **California kingsnake** (*Lampropeltis californiae*) is not seriously affected by rattlesnake venom, and it will feed on rattlesnakes, other snakes, and lizards, as well as birds, eggs, and mammals. The kingsnake is a constrictor, striking and holding its prey with its teeth and then quickly coiling around it and squeezing until the prey cannot breathe. All constrictors kill in this manner, suffocating rather than, as is sometimes believed, crushing their prey.

The kingsnake's color pattern is extremely variable. The California subspecies is typically dark brown or black, with creamy yellow or white bands or irregular patches. Some common kingsnakes are quite docile, and others will defend themselves ferociously: snakes, like people, have unique temperaments.

In the natural world, a common tactic employed by

California kingsnake
(*Lampropeltis californiae*)

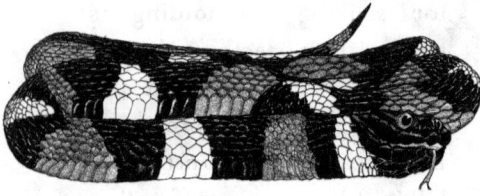

Mountain kingsnake
(*Lampropeltis zonata*)

harmless creatures is to mimic harmful animals. The **gopher snake** (*Pituophis catenifer*) resembles a rattlesnake. Their brown patterns are similar, and when cornered, the gopher snake hisses, flattens its head, and shakes it tail (which bears no rattles), creating a passable imitation of a rattlesnake. To the snake's peril, this bluff sometimes leads people to believe gopher snakes are dangerous, which they are not.

The gopher snake is well named, as rodents comprise the majority of its diet. It will search burrows for squirrels, gophers, young rabbits, and mice and swallow them whole. Gopher snakes are therefore important natural rodent controls, and they will also climb to nests for eggs, prey on birds, and eat lizards. They often kill large prey by constriction.

The buzz of a **western diamondback rattlesnake** (*Crotalus atrox*), the only venomous snake of the Valley/Foothill region, is one serenade of nature that is guaranteed to get your heart pumping. But the truth is that snakebites are quite rare and can be avoided by always looking before placing hands and feet in snake country. Our local rattle-

Gopher snake (*Pituophis catenifer*)

snake is often quite timid and will, if allowed, almost always retreat. It is wasteful for rattlesnakes to use their venom on creatures too large to swallow. Their warning rattles probably evolved to minimize this occurrence. Rattlesnakes are beneficial, but long-term killing of rattlesnakes and habitat loss have diminished local populations of this important form of pest control.

Rattlesnakes sense temperature differences of a fraction of a degree. Heat-sensitive pits near the nostrils, for which the pit viper family is named, help them to locate small mammals, in complete darkness if necessary. Striking rapidly, the rattle-snake injects venom from fold-out, syringe-like fangs. It uses its forked tongue to collect scent chemicals left by its fleeing prey so that, after first allowing it to die, it can track its victim and consume it head first.

Rattles are made of the same material as scales, with a new button being formed at each molt of the skin. Snakes shed their skins irregularly, depending on temperature and the amount of food they have eaten; contrary to popular belief, each button does not represent one year. Rattlesnakes give birth to live young, which are born equipped with venom and one button.

Western diamondback rattlesnake
(*Crotalus atrox*)

BIRDS

Pelicans and Wading Birds | Cormorants | Grebes | Ducks | Rails, Coots, and Cranes | Shorebirds | Turkey Vultures | Raptors | Owls | Quail and Pheasants | Roadrunners | Woodpeckers | Belted Kingfishers | Poorwills | Hummingbirds | Pigeons and Doves | Perching Birds

Bird bodies are modified to meet the seemingly opposing demands of flight: they must be strong and lightweight, rigid and flexible. Their long, hollow bones are light and strong, due to internal bracing, and many bones are fused together, providing a rigid attachment site for powerful muscles. Birds lack heavy teeth and instead have beaks made of lightweight keratin protein. (Chewing is done with a muscular gizzard, a grinding organ near the bird's center of gravity.) Bird bodies are so light that the feathers sometimes outweigh the skeleton. Even the reproductive organs are lightweight, shrinking to almost nothing when birds are not breeding.

BEAK TYPES

A. Omnivore
B. Insectivore
C. Predator
D. Strainer
E. Fish and frog diet
F. Seed eaters
G. Woodpecker

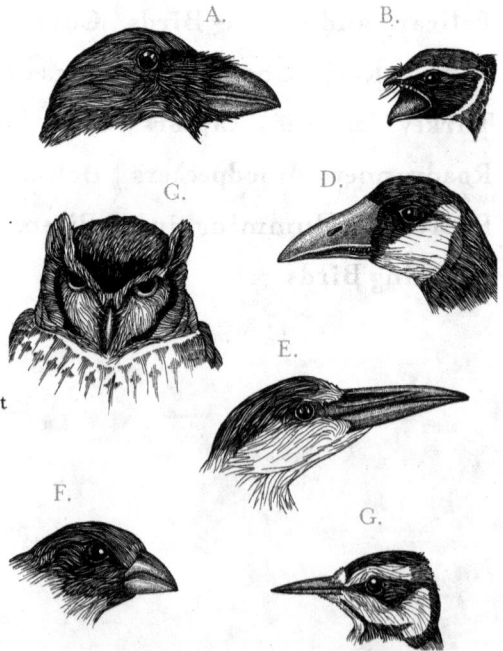

Bird muscles are supported by a high metabolism that results in abundant power but also produces high body temperatures. For example, a house sparrow has a normal body temperature of 107°F. High metabolism requires a large oxygen supply and efficient cooling, which is provided by lungs and air sacs. Birds have at least nine air sacs, some of which are located in the large, hollow bones.

Feathers are an amazing feature. They provide lift as an airfoil and also serve as insulation, protection, camouflage, and a means of communication. Birds maintain their feathers by preening: cleaning each feather by running it through the beak and waterproofing it with oil from a gland located at the base of the tail. Worn-out feathers are molted each year.

A bird's wings are its arms, so the beak and feet must do the job of hands. The typical perching foot has three toes forward and one back: tendons tense as the bird relaxes its weight on its feet, allowing it to sleep on its perch. The talons of birds of prey are similar but stronger, with large claws. Zygodactyl feet have two toes forward and two back. They enable woodpeckers and cuckoos to grasp branches and tree trunks. Webbed or lobed feet are used for paddling in water.

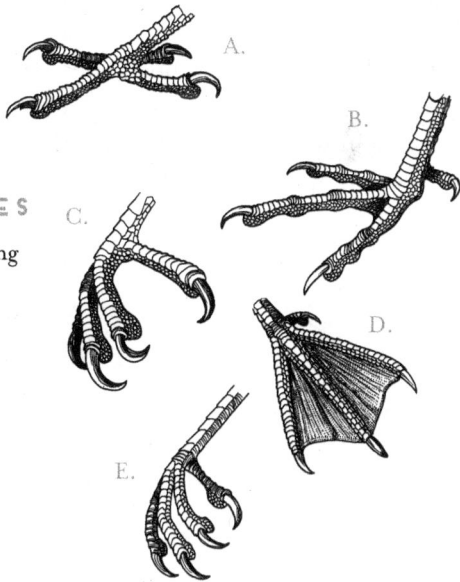

BASIC FOOT TYPES
A. Running or climbing
B. Running
C. Grasping (raptor)
D. Swimming
E. Perching

The mysteries of bird migration have intrigued people for millennia. The ancient Greeks believed that some birds hibernated in holes or beneath water, and that others actually changed into different species. Today it is known that many bird species fly seasonally to different localities, usually to find food or mates. Migrations may be latitudinal, with birds "flying south for the winter," or elevational, with birds moving downslope in winter from the icy mountains. Bird migrations serve as a natural calendar, reflecting the rhythm of seasonal cycles.

The musical ability of birds varies widely, with some remaining practically mute, some using other means of communication, such as drumming on wood, and still others singing beautifully with great gusto. In most species, males sing more than females, as they must defend a territory and attract a mate.

Courting can be hazardous for males because it exposes them to predators, and the sexual event usually takes place in seconds. The tasks of building nests, incubating eggs, and caring for young are performed by the female alone in some species or shared by both parents in others. The extent to which the young are developed at hatching and the amount of parental care required vary from one species to the next. Precocial birds, such as waterfowl and pheasants, are fully feathered, have their eyes open, and are ready to move about almost immediately after hatching. Altricial birds, such as robins and most perching birds, hatch naked with their eyes closed and are capable of only very limited movements.

Pelicans and Wading Birds
(order Pelecaniformes)

Pelicans are spectacular water-birds, especially in flight as they use thermals to gain elevation or to cruise above water surfaces. They fish by dipping their large bills like nets while swimming. **American white pelicans** (*Pelecanus erythrorhynchos*) nest on islands and peninsulas of inland lakes.

Wading birds (family Ardeidae) are long-legged hunters of the shoreline shallows. They use long necks and daggerlike bills to capture a variety of animal life. Variations in body size and in the lengths of legs and bills, as well as varying times of activity, reduce competition in the habitats that the different species of herons share. Herons have several patches of unique feathers known as powder down on their breasts and flanks. Powder down is never molted, and the tips continuously break down into a talclike powder used to condition and waterproof feathers when preening. In flight, herons bend their necks into an "S" shape and trail their long legs. Most of these birds build platform nests in colonies, known as rookeries.

Pelican in flight

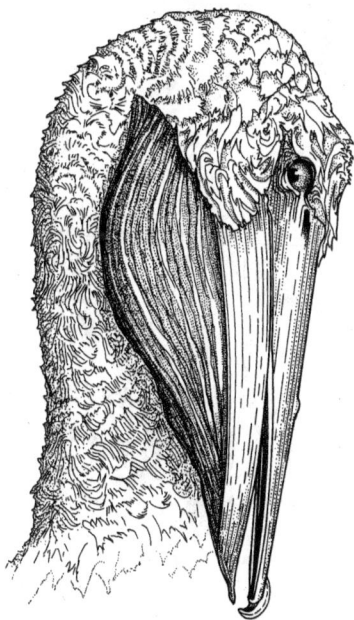

American white pelican
(*Pelecanus erythrorhynchos*)

The **black-crowned night-heron** (*Nycticorax nycticorax*) is nicknamed "night crow." It can be startling to encounter flocks of them roosting in trees. The call, often uttered at night, is a low, nasal "kwak." They hunt alone by the light of the moon, wading along banks in search of small fish, shrimp, and frogs. The **green heron** (*Butorides virescens*), another crow-sized wading bird, also hunts alone.

Snowy egret (*Egretta thula*)

Plume hunters wiped out California populations of egrets in the late 1800s. Now, with a completely protected status, egrets have returned to many of their former haunts. The three species found in the Valley/Foothill region each have a different lifestyle. The introduced **cattle egret** (*Bubulcus ibis*) can often be seen out in pastures following livestock around. These bright white but orange-tinted egrets find hunting best where cattle stir up the grass and scare up food for them. The small, dark bills of **snowy egrets** (*Egretta thula*) are good at snatching up small prey, allowing these birds to avoid competition with the large-billed **great egret** (*Ardea alba*).

Great blue heron (*Ardea herodias*)

Looking like feathered dinosaurs, **great blue herons** (*Ardea herodias*) are often seen flapping slowly overhead or stalking along shorelines and

in fields, capturing prey ranging from fish to young squirrels. Adding to the prehistoric
look is the bird's great size, with
a height of nearly four feet and
close to six feet of wingspan.

American bitterns (*Botaurus lentiginosus*) are brown-and-
white-breasted herons with a
neat camouflage trick. When
threatened, a bittern remains
almost motionless, swaying in
concert with the breeze, with its
bill pointed straight up, blending with the vertical marsh vegetation. If its deception fails,
the bittern erupts upward in a
startling flush of wings.

Cormorants
(order Suliformes)

The bill of the **double-crested
cormorant** (*Phalacrocorax auritus*)
is hooked at the tip to help the
bird grip fish. A heavy body
and a lack of oil in the wings
contribute to the cormorant's
prowess as a diver: one scuba
diver was startled to see a cormorant swim by at sixty feet
below the surface.

Double-crested cormorant
(*Phalacrocorax auritus*)

American bittern (*Botaurus lentiginosus*)

Green heron (*Butorides virescens*) under a black-crowned night-heron (*Nycticorax nycticorax*). Great egret (*Ardea alba*) among rushes behind a Wilson's snipe (*Gallinago delicata*)

Virginia rail (*Rallus limicola*) under a red-winged blackbird (*Agelaius phoeniceus*)

Grebes
(order Podicipediformes)

Cruising low in open water and diving to snatch fish, grebes are loners. Young grebes sometimes ride on their mothers' backs, even during long dives. Grebes are sometimes mistaken for ducks, but no duck has semi-webbed feet or such a pointed bill.

Pied-billed grebe (*Podilymbus podiceps*)

Western grebe (*Aechmophorus occidentalis*)

Ducks
(order Anseriformes)

Web-footed and fleet of wing, ducks do a lot more than just dabble at greens on the local pond. Each fall, millions of ducks make the southward journey to warm climates where food is abundant. Interior California serves as part of the great Pacific Flyway, where ducks stop over to eat and rest before moving onward. During this southward parade, ponds that have been vacant all summer suddenly erupt with the calls of ducks.

The **mallard** (*Anas platyrhynchos*) is one of the few ducks that actually quack. Other species may cackle, whistle, grunt, or erk. Ducks are not all built the same either. Diving ducks have powerful legs toward the backs of their bodies. Such anatomy provides thrust, but it comes at a cost. For example, **ruddy ducks** (*Oxyura jamaicensis*) cannot walk on land and are poor fliers, but they are superior divers.

The beaks of ducks are modified to fit their lifestyles as well. Each species has a different-sized strainer in the beak to filter out plant food and plankton. Species, such as the **common merganser** (*Mergus merganser*), that feed on fish have toothed edges on their beaks to hold on to slippery prey.

COMMON DUCKS

A. Wood duck (*Aix sponsa*): Many wood ducks remain in the Valley year round, where they nest in trees.

B. American wigeon (*Anas americana*): A large, white wing patch shows up when this large duck is in flight.

C. Ring-necked duck (*Aythya collaris*): A vertical white stripe on each side and sometimes a ring on the bill identify this duck.

D. Common merganser (*Mergus merganser*): These great swimming ducks capture fish with their narrow, hooked bills.

E. Ruddy duck (*Oxyura jamaicensis*): These stiff-tailed diving ducks have black caps and white cheeks.

F. Northern pintail (*Anas acuta*): Pintails are large-bodied ducks with long necks, gray bills, and pointed tails.

G. Canvasback (*Aythya valisineria*): Males have brilliant white backs and black breasts and tails.

H. Cinnamon teal (*Anas cyanoptera*): This small, rust-colored dabbling duck tips bottoms-up when feeding near the water's surface.

I. Common goldeneye (*Bucephala clangula*): Wings make a whistling sound. A male goldeneye has a large, white spot before each eye.

J. Bufflehead (*Bucephala albeola*): This is a small diving duck with a white eye-stripe and a distinctive crest.

F.

G.

H.

I.

J.

Rails, Coots, and Cranes
(order Gruiformes)

Most of the Valley/Foothill rails look at first like scared chickens that live alone in swamps. But look at the beak, and those enormous feet! Long toes spread out the body weight, much like snowshoes do, allowing rails to stroll around on muddy banks without sinking up to their rumps. Local rails are secretive birds, except for coots, which often come ashore and noisily feed on plants. Each rail species has a diet relating to the shape of its beak. **Virginia rails** (*Rallus limicola*) use a long, probing beak for fishing around deep in mud for small shrimps, while the shorter-beaked **sora** (*Porzana carolina*) skims food closer to the surface.

The great v-shaped flocks of enormous wading birds that pass overhead in the Central Valley each winter are **sandhill cranes** (*Antigone canadensis*). Sandhill cranes breed in the Arctic. In California, they overwinter in a few locales, including the Sacramento–San Joaquin Delta, Merced County, and the Carrizo Plain. Unlike herons, cranes fly with their necks outstretched. They are known for their bizarre courtship dance, in which they lower their heads, flap their wings, and hop about.

Sora (*Porzana carolina*)

American avocet
(*Recurvirostra americana*)

American coot (*Fulica americana*)

Shorebirds
(order Charadriiformes)

In one of nature's best examples of niche partitioning, shorebirds show just how many similar species can exist together. They seem to all feed the same way, striding or dashing along beaches, probing the sand with their beaks for a meal. But how they do it and the tools they use differ. **American avocets** (*Recurvirostra americana*) swing their upturned bills from side to side while feeding. **Black-necked stilts** (*Himantopus mexicanus*) have such long legs that, rather than bend them to feed on land, they wade into water to forage. Each species of sandpiper has a slightly differ-

Killdeer (*Charadrius vociferus*)

ent beak and selects foods that similar species miss.

Some shorebirds exhibit interesting tactics to distract enemies from their eggs. The parent **killdeer** (*Charadrius vociferus*) acts as though its wing is broken, while calling and moving away from the nest, enticing the intruder to follow. When it has put enough distance between the nest and its enemy, the killdeer "recovers" and flies away, repeating its ruse if the nest is again approached.

Not all gulls are "sea gulls": **California gulls** (*Larus californicus*) are often seen in school yards and in fields, following tractors. Flocks of gulls often appear inland before a major storm hits the Pacific. **Terns**

Black-necked stilt (*Himantopus mexicanus*)

California gull (*Larus californicus*)
Caspian tern (*Hydroprogne caspia*)

are streamlined relatives of gulls that feed on fish, which they typically catch with a dramatic headfirst dive.

Turkey Vultures
(order Cathartiformes)

Soaring in a slow, wheeling flight, **turkey vultures** (*Cathartes aura*) combine keen senses of smell and vision to locate dead animals from great distances. Not only do they search the ground for food, but turkey vultures keep an eye on each other. When a circling vulture descends to eat, it makes a rocking motion with its wings that other vultures notice from miles away. Soon the sky is swirling with hungry birds. The turkey vulture's featherless head, which looks something like a turkey's head, seems too small for its large black body, but it is that bald head that allows a vulture to retrieve tidbits from within dead carcasses without messing up its feathers. Rather than carry their food off to eat elsewhere, turkey vultures are likely to gorge themselves on the spot until they are almost too heavy to fly. This behavior makes sense, considering that the next meal may not come for days.

Turkey vulture (*Cathartes aura*)

Raptors
(order Accipitriformes)

Many birds kill other animals, but raptors are among the few that use claws instead of beaks to snatch their prey. The raptor's beak is curved into a hooked tip that penetrates flesh to make a kill, and the talons of most raptors are so robust and curved that they work like meat hooks to carry a victim off. Different species of raptors hunt everything from grasshoppers to rabbits. **Bald eagles** (*Haliaeetus leucocephalus*) are among the few raptors that include dead animals in their diet. Male and female raptors look similar in many species, but often the female is larger. These eagles are winter visitors to many reservoirs but return north to lay eggs and raise their young.

Golden eagle
(*Aquila chrysaetos*) in flight

Red-shouldered hawks (*Buteo lineatus*) patrol the sky along rivers, taking crayfish, rabbits, and whatever else they can catch. These hawks rarely hunt when soaring; instead, they make low-flying forays from a perch. Their call is a series of four to five high, urgent screams. They are darkly checkered with black and white on the backs of their wings. The breast is barred with rust. Flight pattern is several stiff wingbeats followed by a glide.

Bald eagle (*Haliaeetus leucocephalus*)

Red-tailed hawks (*Buteo jamaicensis*)

Red-tailed hawks (*Buteo jamaicensis*) are the birds often seen wheeling overhead, emitting a single scream with gusto. These robust hawks eat anything from lizards to rabbits, capturing their prey in a steep dive from a vantage point high in the air. Among their favorite perches are fence posts, where they watch for rodents flushed out by cars. They display several color forms, depending on their age and location, but all have a rust-colored tail when mature.

Cooper's hawk (*Accipiter cooperii*)

The **Cooper's hawk** (*Accipiter cooperii*), formerly called a chicken hawk, are brown-streaked birds with long, banded tails. These hawks fearlessly pursue prey at high speed through dense cover, even walking or diving into shrubbery to grab a cowering bird. Small birds make up about 90 percent of the diet of this little hunter. Using its large talons to kill its prey, the Cooper's hawk carries it to a branch within a leafy canopy, where it plucks it of its feathers and devours it. The **sharp-shinned hawk** (*Accipiter striatus*) is a smaller version of the Cooper's.

Red-shouldered hawk (*Buteo lineatus*)

RAPTORS

A. Swainson's hawks (*Buteo swainsoni*) have dark heads and chests and banded tails. They migrate to South America to overwinter, often in groups of over fifty.

B. Ferruginous hawks (*Buteo regalis*) have dark legs and wingtips. These large raptors are winter visitors here.

C. Northern harriers (*Circus cyaneus*) can be identified by their low, swinging flight and white rump patch. Males perform aerial loops and rolls during breeding season. Their grass nests are placed on the ground or in low thickets.

D. Prairie falcons (*Falco mexicanus*) are tan, with pointed wingtips and a dark eye bar. Built somewhat like jets, they fly incredibly fast.

E. White-tailed kites (*Elanus leucurus*) sometimes hover. Their descent upon prey looks more like parachuting than diving, as they do not fold their wings.

During the great grasshopper hatch in spring, the **American kestrel** (*Falco sparverius*), our smallest and most common falcon, is often seen perched on fence lines or hovering over fields. These falcons pounce on hundreds of the little hoppers each day. They carry some off and eat them at their perches or feed them to their chicks. Later in the season, kestrels track rodents by following their almost invisible urine trails. Like many other raptors, kestrels often cache leftover food to be eaten at another time.

American kestrel (*Falco sparverius*)

Owls (order Strigiformes)

These raptors are great nocturnal stealth hunters. Because their faces are disc-shaped, their ears are placed in a way that enhances their hearing and allows them to detect the faintest rustles of their prey. One ear opening is higher than the other, for still more accuracy of hearing. Soft feathers dampen whistling noises as owls fly, and an outside toe points forward or backward for snatching prey. Because it has only one condyle (bony projection) on the skull, an owl can turn its head past its shoulder as it watches and listens. Unlike hawks, which rend their prey, owls swallow small prey whole.

If owls come out during the day, they are likely to be mobbed by perching birds. To keep unwanted visitors away, young owls make a call that resembles the buzz of a rattlesnake. The spine-chilling screech that many owls make at night causes their victims to panic and make a mad dash from their hiding places.

Great horned owls (*Bubo virginianus*) are powerful hunters, capable of killing skunks, cats, and even sleeping hawks. Often, you will hear their deep "who-who" without actually seeing an owl. They may emerge and perch before dark, seeming to take their time waking up before hunting.

While hunting in the night sky, **western screech owls** (*Megascops kennicottii*) make single clucking sounds. They also make a soft and mellow series of low-pitched whistles often heard at night, even in cities.

Burrowing owls (*Athene cunicularia*) nest in squirrel holes and other abandoned burrows and perch conspicuously on the ground or on low objects during the day. When alarmed, they bob up and down on their long legs.

Common barn owls (*Tyto alba*) hunt primarily by sound, preying on small rodents. They roost by day in old buildings, trees, and other dark places. Eerie, raspy screeches and clicking sounds are their most often heard vocalizations. Barn owls fly low and are blinded by headlights; many are struck by cars.

A.

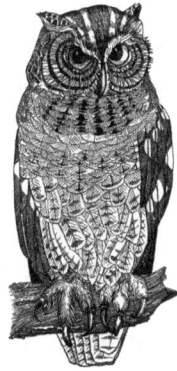

B.

C.

D.

OWLS

A. Great horned owl (*Bubo virginianus*)
B. Western screech owl (*Megascops kennicottii*)
C. Burrowing owl (*Athene cunicularia*)
D. Barn owl (*Tyto alba*)

Quail and Pheasants
(order Galliformes)

Running birds, which include quail and pheasants among their numbers, have great circulation in their leg muscles, with an ample supply of oxygen-rich blood. This is why "dark meat" is in the legs and the lighter meat (with a reduced blood supply) is in the flight muscles of the breast.

California quail (*Callipepla californica*) seem not so much to run as to roll along the ground as they dash for cover in bushy thickets. They forage for seeds, tender leaves, and similar foods in open areas near shelter. The call of the male is an emphatic "chi-ca-go." Their alarm call is a high, excited "whip-whip," and a similar call reassembles the flock.

Ring-necked pheasants (*Phasianus colchicus*), which have a diet and lifestyle similar to the quail's, were brought here from Asia for sport hunting. Males compete for a harem of females; a single male may mate with ten to twenty females. Its call is a two-syllable squawk.

Ring-necked pheasant
(*Phasianus colchicus*)

Male and female California quail
(*Callipepla californica*)

Roadrunners
(order Cuculiformes)

Perhaps the most remarkable running bird of all, the **greater roadrunner**, (*Geococcyx californianus*) is zygodactyl: its two toes pointing forward and two toes backward leave x-like tracks. These speedy birds inhabit dry habitats, where they are aggressive hunters of insects, reptiles, and rodents. They earned their name by running in front of horse-drawn carriages and have been clocked at fifteen miles per hour. To save energy, they allow their body temperature to drop at night and use the morning sun to warm up again, much like reptiles.

Woodpeckers
(order Piciformes)

Not all woodpeckers hammer on wood, but all true woodpeckers do have a chisel-like beak and a stiff tail with which to brace the body against tree trunks. Their zygodactyl foot configuration is handy for climbing. They have incredibly long and sticky tongues, which many use for slurping up insects.

The **acorn woodpecker** (*Melanerpes formicivorus*) is a clown of a woodpecker with black with white wing patches that makes

Greater roadrunner
(*Geococcyx californianus*)

Acorn woodpecker
(*Melanerpes formicivorus*)

Bewick's wren (*Thryomanes bewickii*), Anna's hummingbird (*Calypte anna*), Cooper's hawk (*Accipiter cooperii*) with American bushtit (*Psaltriparus minimus*), evening grosbeak behind leaves (*Coccothraustes vespertinus*). Nuttall's woodpeckers (*Dryobates nuttallii*, top left) eat insects that bore into tree bark.

Hairy woodpecker (*Dryobates villosus*, top left). The northern flicker (*Colaptes auratus*) has almond-colored wing linings and a white rump. It forages for seeds on the ground and is one of the few animals in the world that can tolerate the taste of ants.

a hysterical-sounding "wacka wacka" call. These social birds also have some interesting behaviors. Only some of the birds breed, while others are helpers that provide food, instinctively collecting and storing acorns and occasional pine nuts in holes drilled in a single tree, known as a granary. Insects, including ants as well as flying insects captured in the air, are an important spring and summer food, and the granary provides food for the winter and supplies energy in spring to helpers and parents that catch insects to feed hatchlings. The granary, with thousands of holes (one was estimated at fifty thousand), is a valuable resource, as each bird can only drill a few holes each year. The urge to fill a vacant hole with an acorn is instinctive.

Belted Kingfishers
(order Coraciiformes)

Belted kingfishers (*Megaceryle alcyon*) patrol waters throughout California's interior. The rattling sound of a kingfisher's call often precedes the sight of the jerky wingbeats that carry this large-headed bird along. Kingfishers hunt from perches overlooking water, and they hover above the surface at times. When they sight their prey, they dive headfirst into the water to snatch it.

Belted kingfisher (*Megaceryle alcyon*)

After slapping its catch, often a struggling fish, against a perch, the kingfisher swallows it headfirst. Families of young kingfishers can sometimes be seen emerging from holes in riverbanks, where they nest. The young learn to fish by retrieving dead or wounded fish.

Belted kingfisher (*Megaceryle alcyon*) and violet-green swallow (*Tachycineta thalassina*)

Poorwills
(order Caprimulgiformes)

In winter, the body tempera-ture of a **common poorwill** (*Phalaenoptilus nuttallii*) drops as much as forty degrees, the breathing rate drops drastically, and the digestive processes stop. Metabolic activities increase as the warm days of spring arrive. Scientists initially dismissed Native American accounts of the

Common poorwill (*Phalaenoptilus nuttallii*) in foreground

poorwill's odd winter torpor as absurd, but later, as more studies were made, credence was given to the early stories. Poorwills make spectacular swooping dives in the evening when hunting insects. At night they sometimes perch on warm asphalt roads.

Hummingbirds
(order Apodiformes)

These birds are named for the humming buzz of their wings, which beat seventy-five times per second in normal flight, slowing to fifty-five beats per second when hovering. Their mode of flight generates power on the upstroke as well as the downstroke.

These active birds feed on flower nectar collected with their long bills and tubelike tongues while hovering. Tiny insects and pollen in nectar supply much of their protein requirement. Hummingbirds establish feeding territories and defend them, which some-times results in dramatic chases around a single hummingbird feeder.

Males often court females with an impressive pendulum

Anna's hummingbird (*Calypte anna*) leaves the garden at sunset, and the western toad begins its nightly patrol.

flight. After breeding, the male leaves the tasks of building a nest and rearing the young solely to the female. The little nest, sometimes reinforced with spider silk, is only one inch wide, and the eggs are the size of pinto beans.

Anna's hummingbird (*Calypte anna*)

Males have scintillating iridescent patches on their heads and throats, instantly dulled by the turn of the bird's head. This is because hummingbird brilliance derives not from bright pigments but from light reflecting off of feather structures.

Pigeons and Doves
(order Columbiformes)

Mourning doves (*Zenaida macroura*) are named for their mournful, cooing call. Farming and other disturbances to Valley and Foothill terrain, as well as the introduction of exotic weeds, have changed the environment in a way that is beneficial to these seed-eating birds. Although they now have more seeds than ever to choose from, they must still face the challenge of finding water on a daily basis.

Doves' nests are notoriously ill-constructed, but both parents contribute to the upbringing of the young—two squabs—feeding them "crop milk," a secretion from the parents' crop walls. A relative, the **rock pigeon** (*Columba livia*), or domestic pigeon, plagues urban areas.

Mourning dove (*Zenaida macroura*)

Perching Birds
(order Passeriformes)

From bold ravens to meek swallows, all true perching birds have feet with three toes forward and one back. Many perching birds are ecological generalists; their niche is broad enough to include a wide variety of foods and habitats. This is exemplified by the presence of jays, crows, sparrows, and their relatives in orchards, along highways, in backyards, and nearly anywhere else where a meal is to be had.

Flycatchers
(family tyrannidae)

During the spring breeding season, **western kingbirds** (*Tyrannus verticalis*) attack any raptor or crow that enters their territory, using their superior maneuverability to avoid counterattacks. Their pugnacious and successful territoriality no doubt gave them their name. For food, kingbirds pursue flying insects. They build their nests in trees, and the parents seem to vocalize constantly in their comings and goings, with a varied, shrill twittering. These birds fly to South America to overwinter and return to California near the end of March.

The **black phoebe** (*Sayornis nigricans*), another flycatcher, is similar in size to the kingbird. Phoebes hunt along rivers. The quiet solitude of a shady pool of water near a bridge is often accentuated by the insistent call of this small, black-headed bird. Phoebes do not migrate, although they may wander locally to find insect prey during winter. Their cup-like nests of mud pellets and vegetation can be found stuck in high places near water.

Black phoebe (*Sayornis nigricans*)

Swallows
(FAMILY HIRUNDINIDAE)

Swallows can often be seen diving gracefully under bridges or sailing along in twittering flocks near water. Strong, acrobatic fliers with wide mouths, they feed on flying insects. Swallows migrate south to avoid winter

Cliff swallow (*Petrochelidon pyrrhonota*)

and follow the warm summer climate in which their insect prey thrives. **Cliff swallows** (*Petrochelidon pyrrhonota*) build mud-pellet nests with narrow necks, often under bridges or the eaves of buildings. **Barn swallows'** (*Hirundo rustica*) cup-shaped mud nests are often found stuck under bridges, in barns, or in irrigation pipes. The **tree swallow** (*Tachycineta bicolor*) and **violet-green swallow** (*Tachycineta thalassina*) are iridescent blue-green above and white below. Both nest in tree hollows and other cavities.

Jays and Crows
(FAMILY CORVIDAE)

These bold birds are despised because they destroy not only crops but the eggs and young of other bird species; their role of nest predator makes other birds regard them with suspicion. Despite these unattractive habits, crows and jays are interesting by virtue of their social interactions. **California scrub jays** (*Aphelocoma californica*) often serve as self-appointed sentinels, greeting visitors with a raucous, raspy "shreep-shreep."

Both jays and crows have beaks designed to process a variety of foods, from nuts and berries to other animals. Jays unintentionally plant many

California scrub jay
(*Aphelocoma californica*)

oak trees by hiding more acorns than they retrieve. About 30 percent of the scrub jay's diet is animal matter, including insects, lizards, mice, bird eggs, and nestlings. Scrub jays have been reported to remove ticks from cattle and deer. This broad diet, an impressive intelligence, and robust body size help crows and jays to dominate nearly everywhere, from backyard to wilderness.

Yellow-billed magpies (*Pica nuttalli*), the flamboyantly long-tailed, black-and-white birds

American crow (*Corvus brachyrhynchos*)

that herald one's presence in the Valley more than any other, have a similar ecology to their close relative, the scrub jay.

Much effort has been made to suppress **American crow** (*Corvus brachyrhynchos*) populations, but they continue to thrive and have taken up residence in small towns throughout the Valley and elsewhere where they were formerly absent. Their call is a series of caws. **Common ravens** (*Corvus corax*), larger than crows, can be identified by their wedge-shaped tails and deep, croaking calls, as well as a peculiar, resonant clicking.

Yellow-billed magpie (*Pica nuttalli*)

Common raven (*Corvus corax*)

BLACKBIRDS, COWBIRDS, AND MEADOWLARKS (FAMILY ICTERIDAE)

The screeching trill of **red-winged blackbirds** (*Agelaius phoeniceus*) is heard in marshes and grasslands the entire length of California. These beautiful birds nest in colonies located among upright stalks of cattails, tules, or grain crops.

Brewer's blackbirds (*Euphagus cyanocephalus*) have adapted superbly to landscapes altered by development, becoming widespread in both town and country despite the fact that only half their young usually survive. Many are eaten by crows, jays, and snakes, with others falling from their flimsy nests.

Brewer's blackbird
(*Euphagus cyanocephalus*)

The metallic yellow color of its eyes helps distinguish the male Brewer's from the **brown-headed cowbird** (*Molothrus ater*). Cowbirds lay their eggs in the nests of other songbird species. Their young develop quickly and push the original songbird eggs or young from the nest.

The **western meadowlark**'s (*Sturnella neglecta*) enthusiastic, melodious call of "two-tea two-doodle-tea" often greets visitors to grasslands. When singing, meadowlarks expose their bright yellow breasts, but they turn their brown camouflaged backs and run when intruders get too close. This countershading occurs in many other animals and results in a slick disappear-

Western meadowlark
(*Sturnella neglecta*)

ing act when situations turn dangerous. Meadowlarks build their nests on the ground, concealed by grassy, domed covers, and eat mostly seeds and insects.

New World Sparrows (family passerellidae)

The sparrows and towhees, busy little birds, all have a basic cone-shaped beak for cracking seeds. Each species of sparrow has a distinctively shaped beak that enables it to specialize on foods of specific dimensions. Thus, large flocks of different sparrow species can forage together and avoid intense competition for single types of food items. Towhees often search for seeds by scratching up leafy litter. The **Savannah sparrow** (*Passerculus sandwichensis*) is an open-country bird with a brown-streaked breast. The **lark sparrow** (*Chondestes grammacus*) has a chestnut cheek and a spot on the chest. The **white-crowned sparrow** (*Zonotrichia leucophrys*) is a winter visitor with black and white stripes on its crown. The **spotted towhee** (*Pipilo maculatus*) has a black head, a long tail, and reddish sides.

Lark sparrow (*Chondestes grammacus*)

White-crowned sparrow
(*Zonotrichia leucophrys*)

Spotted towhee (*Pipilo maculatus*)

Savannah sparrow (*Passerculus sandwichensis*) perched above a wrentit (*Chamaea fasciata*)

Old World Sparrows
(family Passeridae)

House sparrows (*Passer domesticus*) were introduced from Europe. They are actually weavers, not true sparrows, a fact reflected in their bulky, enclosed nests, which are so unlike the neat, open cuplike nests that true sparrows build. House sparrows feed in flocks, typically on food provided by human society, and are common to town and country. The males have black bibs.

House sparrow (*Passer domesticus*)

Finches
(family Fringillidae)

House finches (*Haemorhous mexicanus*) and a few other finches show an interesting color change related to food and breeding. In early spring, males feed heavily on plants containing brilliant orange and yellow pigments. These carotenoids appear in new feathers, changing a drab male into a bright beacon for attracting females.

Goldfinches (*Carduelis* spp.) are tiny yellowish birds with black caps that migrate down from the Sierra slopes. They don't come to breed, but to scratch for seeds on our lawns until warm days return.

Goldfinch (*Carduelis* spp.)

Brown Creepers
(family Certhiidae)

During the winter, many **brown creepers** (*Certhia americana*) descend from the icy

Sierra Nevada to milder climates, where they forage on the bark of trees. Creepers can be spotted by their distinctive way of climbing tree trunks, spiraling upward and using their curved bills to snatch spiders and insects that live in the crannies of furrowed bark.

EUROPEAN STARLINGS (FAMILY STURNIDAE)

Starlings (*Sturnus vulgaris*) were introduced from Europe in 1890 and spread like a plague, coming to California in the 1940s. Flocks of over fifty thousand starlings have since been seen over Valley fields, destroying crops. Further damaging their already poor

Brown creeper (*Certhia americana*)

Starling (*Sturnus vulgaris*)

reputation is their habit of driving off native birds that compete for the same nesting cavities they prefer. Starlings are aggressive, often squabbling and pecking each other, which is why you will see them at least a beak's length from each other on perches. Their calls are varied, consisting of a chaotic series of clicks and squeaks as well as imitations of other birds' calls.

Cedar Waxwings
(family bombycillidae)

Waxwings, so named for the waxy structures on some of their feathers, are migrants, stopping in as they move south to seek winter food and then again on their way north to breed. They sometimes become drunk on vine-fermented berries.

Northern Mockingbirds
(family mimidae)

The song of a **northern mockingbird** (*Mimus polyglottos*) is melodious and clear, and typically delivered from a visible perch, such as an antenna or power pole. They are named for their habit of imitating, or mocking, the songs of other birds, adding these borrowed songs to their own impressive medley. Males do most of the singing, and unmated males often sing for most of the night, hoping to attract a mate and establish a territory. Mockingbirds are so aggressively territorial that they will even attack their own reflections in hubcaps.

Cedar waxwing (*Bombycilla cedrorum*)

Northern mockingbird
(*Mimus polyglottos*)

LOGGERHEAD SHRIKES (FAMILY LANIDAE)

Shrikes are songbirds that have somewhat adopted the hunting techniques of hawks. A notch near the curved tip of a shrike's beak allows it to dislocate the necks of its prey. Shrikes eat rodents, lizards, small birds, and large insects. Because prey that is not eaten on the spot is impaled upon a thorn or fence barb, as seen in the illustration below of a **loggerhead shrike** (*Lanius ludovicianus*) with grasshoppers it has skewered on branches, shrikes have earned the nickname "butcher bird." Besides food storage, this behavior may also serve to mark territory and advertise hunting prowess to potential mates. Or perhaps shrikes prefer meals that are no longer twitching.

Western kingbird (*Tyrannus verticalis*) in flight with loggerhead shrike (*Lanius ludovicianus*)

MAMMALS

Marsupials | Shrews and Moles | Bats | Rabbits and
Hares | Rodents | Carnivores | Hoofed Mammals

Though they are conspicuous by virtue of their ecological importance, size, and behavior, mammals can be difficult to observe in a natural setting. This is because many mammals are nocturnal, sleeping during the day, when humans are most active. Also, like the pocket gopher, who lives underground, and the beaver, who lives in the water, many mammals occupy habitats where they are not readily seen.

Mammals use a larger proportion of their brains for memory and learning than any other group of animals. Because development of the brain and accumulation of knowledge occur slowly, mammals need more time for embryo development and more parental care than other animals.

Most mammals rely upon their sense of smell as the primary means of sensing their environment, and many have scent glands that are used for communication. (Glands of various types—sweat, oil, scent, and mammary—are characteristic of mammals.) Nonetheless, keen hearing and vision are found among some mammals.

Teeth reveal much about the diet and ecology of a mammal. The incisors, in front, are good for clipping vegetation; canine teeth, used for stabbing or biting, are like cone-shaped steak knives; behind the canines are the premolars, which can produce a shearing action for slicing meat; and farther back in the mouth are molars, for grinding. These can have both flattened and sharp crowns (run your tongue across your molars to experience this).

Mammals that have all four types of teeth have some choice in what they eat; bears, raccoons, dogs, people, and many other mammals have an omnivorous diet. In fall, many omnivores feast on ripening fruits, using their large molars to crush hard seeds and fruit rinds. The flush of insects and rodents in spring causes many omnivores to switch to a predatory lifestyle, using their canines to make the kill. The shearing action of premolars slices meat into chunks that can be swallowed. Cats have no true molars and are among the few mammal groups that rely totally on meat. Not unlike cats in terms of diet are badgers and several other members of the weasel family.

Lacking the penetrating canines of carnivores, herbivores rely on the clipping action of incisors and the crushing ability of molars to process plants. A few rodents make the switch to a carnivore lifestyle by using their incisors to clamp down on insects. A steady diet of coarse plant foods wears down the teeth and can eventually end the life of an old, toothless herbivore. Rodents and horses have ever-growing molars that resist the erosional destruction of an herbivore's diet.

MAJOR TOOTH GROUPS

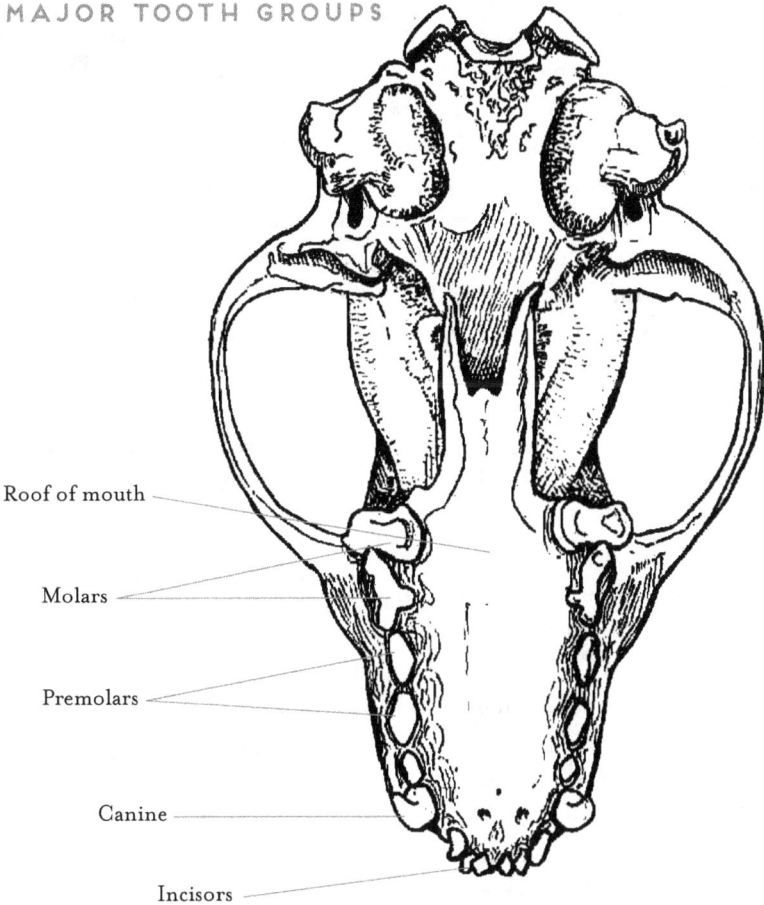

Roof of mouth

Molars

Premolars

Canine

Incisors

Hair, a feature unique to mammals, serves a lot of purposes: insulation, camouflage, and protection from such dangers as sunburn and insect attacks, to name a few. Unlike the hair on a fuzzy insect, mammal hair is composed of keratin protein that arises from hair follicles deep in the skin. Many animals grow or shed hair seasonally and groom it assiduously to provide optimal insulation. Otters, for example, spend several hours each day grooming their fur to insulate themselves from cold water. Drab-colored hair and cryptic markings, together with the fact that most mammals are color-blind, allow most mammals to protect themselves by blending in with their surroundings.

Because they have an internal metabolic furnace and hair for insulation, mammals can adjust their behavior to maintain a fairly constant body temperature and be active in a variety of climates. An active mammal can increase its metabolic rate and heat production as much as tenfold. To cool off, it releases moisture by evaporative cooling: sweating or panting. If it is too cold, a mammal may increase the insulating effect of fur by fluffing it up to trap more air. Shivering contracts the muscles, producing more heat and increasing body temperature.

Evaporative cooling and shivering both require greater-than-normal use of energy, so mammals tend to avoid extreme temperature conditions by retreating to shelter, huddling with other mammals to decrease the amount of surface area exposed, or pressing their bodies to the cool or warm earth. Migration is a more extreme behavior in which a mammal travels a long distance (and seasonally returns), typically to a different elevation or latitude. Besides a more favorable climate, migration often offers a better supply of food. But because it takes so much energy, migration is rare among mammals.

Torpor, the result of decreasing energy use by reducing metabolic rate, is accompanied by a drop in body temperature, heart rate, and breathing rate. Small mammals—some bats and shrews, for instance— do this nightly. Long-term seasonal torpor to avoid cold is known as hibernation. When torpor is used to avoid unfavorable summer conditions, it is known as estivation.

Mammals possess other, less readily visible distinguishing features. The unique muscle that separates a mammal's chest and abdominal cavities is known as the diaphragm. By contributing to mammals' ability to breathe efficiently, this muscle facilitates temperature regulation. Another unique mammalian feature is the palate that separates air and food routes, allowing mammals to breathe easily while eating.

The majority of mammals, including mice, dogs, and deer, give birth after a relatively lengthy gestation period in which the developing young receive nourishment from the mother by an organ of exchange called the placenta. A few marsupials complete their development after birth, in their mothers' pouches.

Mammary glands enable the mother to continue foraging for food and convert it to a nourishing fluid containing protein and fat, the perfect baby food. Milk varies in composition from species to species, and may even change in composition during the development of the young within a species.

Many mammals have a relatively long developmental period during which parents teach their offspring skills to increase their chances of survival. Species that are intelligent and have fewer offspring guard their young with great ferocity, while mammals that give birth to several large litters each year, like rodents and rabbits, provide less care and defense.

Mammal tracks reveal the many forms that the mammalian foot can take. In a plantigrade foot posture, the entire foot carries the weight of the animal. Such a pattern is adapted for a walking gait, as is found in bears, raccoons, and primates. Mammals with a running or stalking lifestyle may have a digitigrade foot posture. In this configuration, found in dogs and cats, the foot bones are lengthened and the body weight rests on the toes. The ankle or wrist joint is so high up on the limb that it's easy to mistake it for an oddly angled knee or elbow. The body weight of powerful runners, such as horses and other ungulates, is borne on toe tips covered by hooves.

When you encounter a set of tracks and you don't know what creature left them, consider the habitat. Raccoon tracks are found in the mud along streams, while jackrabbit tracks are most likely on a dusty path. Toe and claw marks also provide clues. Cats, for example, retract their claws while walking, so they leave no visible claw prints.

Marsupials
(order Didelphimorphia)

Marsupials are unique, in that the tiny young are born in an underdeveloped state—thirteen days after conception in the **common opossum** (*Didelphis virginiana*)—then crawl to their mother's pouch, called the marsupium, where they complete their development after attaching to the teat of a mammary gland. Marsupials have smaller brains, relative to body size, than the placental mammals.

Common opossums have grayish-white fur, pink noses, dark ears with white tips, and long, scaly, prehensile tails. They have fifty teeth, more than any other North American mammal. Originally found in the southeastern United States, these cat-sized animals have become well established in the Pacific states. Resembling a huge rat, the opossum furthers its intimidating appearance with an open-mouthed, very toothy threat display accompanied by hissing and salivating. If it fails to scare intruders, an opossum may resort to "playing possum"—feigning death.

Opossums eat both meat and plant foods, including fruit, garbage, insects, and carrion. This omnivorous diet has allowed them to expand their range to human habitats, where they may be seen after dusk walking across fence tops in the suburbs or on country roads. Opossums are prolific breeders, which compensates for their high mortality rate. They locate their dens in hollow trees, under buildings, or in abandoned burrows, sometimes using their tails to carry vegetation to line their dens. The odd possum will sometimes hang from a branch by its tail.

Common opossum
(*Didelphis virginiana*)

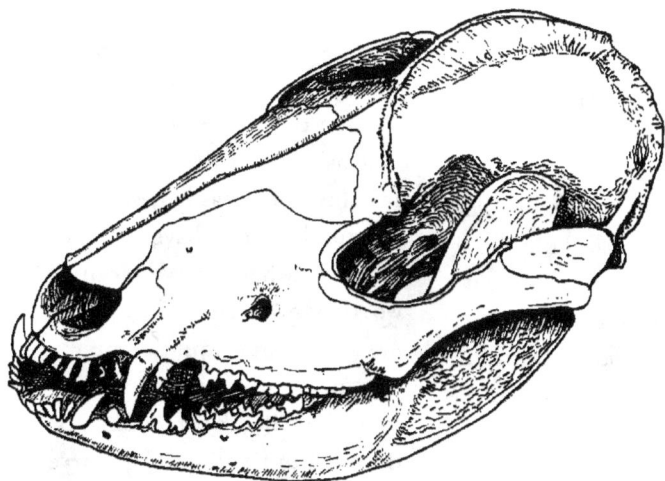

Opossums are very toothy, with over twenty-two teeth in the upper jaw.

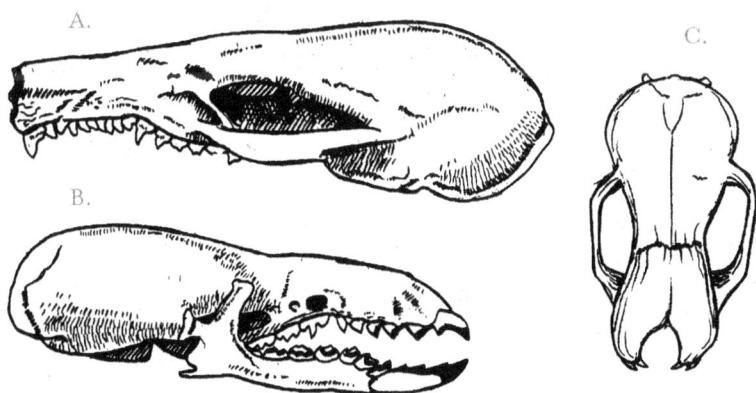

Insectivores: Moles and shrews are tiny mammals with skulls usually less than two inches (5 cm) long. Their incisors are sharp and long.

OMNIVORE AND INSECTIVORE SKULLS

A. Mole
B. Shrew
C. The tiny skull of an insect-eating bat is less than an inch (2.5 cm) long, and there is a notch above the mouth.

Eulipotyphlans

(order Eulipotyphla)

Eulipotyphlans are secretive mammals with small eyes and ears, fine fur lacking long guard hairs, and five toes with claws on each foot. The **ornate shrew** (*Sorex ornatus*) is found in the Foothills and, rarely, in the Valley. Shrews are tiny, some weighing less than one ounce. Such small bodies are greatly affected by temperature variation. Just as a shallow puddle rapidly heats and cools, so the shrew's body temperature would fluctuate if not for its ability to thermoregulate; an almost constant intake of calories to fuel this small dynamo's physiological furnace helps the shrew to survive its size handicap. Hugging the shadows, shrews use their red, needle-like teeth and sometimes paralyzing venom in their saliva to subdue their prey. Every move is a balance of life and death, for shrews die easily from starvation, but they have a list of enemies longer than a month's worth of groceries. The choices are simple but would be unappealing to us: hunt and eventually be killed by an enemy, or do not hunt and quickly starve to death.

The **broad-footed mole** (*Scapanus latimanus*), a common but rarely seen Valley mole, has a naked muzzle and paddle-like hands for digging. Earthworms, insects, and plant material, located by smell and obtained by digging, comprise the majority of a mole's diet. Molehills, often confused with gopher mounds, normally contain clods of earth and have a central opening covered by an eruption of soil pushed up by the mole. Gopher mounds are fan-shaped , with an off-center burrow and often a soil plug.

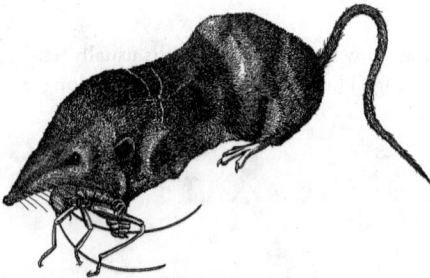

Ornate shrew (*Sorex ornatus*)

The world of a broad-footed mole (*Scapanus latimanus*)

Bats (order Chiroptera)

Bats are the only mammals capable of true flight. Their wings provide power and maneuverability. Except for thumbs, the bones of their arms and hands are long, with flight membranes extending between them and the body, legs, and tail. Movement at the elbow and wrist is restricted to one plane. The shoulder joint is braced with muscles providing the power and rotational support needed for changing direction while flying. Tapering wingtips reduce the buildup of kinetic energy and allow precision in high-speed maneuvers.

Recent scientific study has revealed the ecological importance of bats, especially in insect control. Some 853 species of bats comprise the second largest order of mammals, exceeded in number of species only by the rodents. All bats in the Valley/Foothill region are insectivorous, catching insects as well as drinking while in flight. Some species consume nearly their own weight in prey during a single night.

Identification is difficult because bats fly at night, establish hidden roosts, and must not be handled. At least seventeen different species of bats, some of which are migratory visitors for a few months each year, occur in this region. These local bats sense the world primarily with sonar, emitting from their vocal cords brief blasts of high-frequency sound that bounce off objects and back to the bat. Most of their transmissions cannot be detected with human hearing, but with luck, you may glimpse a bat and hear the overhead clicks of its echolocation in the evening sky.

Thick sinuses, fat, and connective tissue serve to prevent bats from being deafened by their own powerful sonic blasts, and for further protection, the muscles of the bat's sensitive inner ear contract when it is emitting sound. A bat can discern differences in the returning sound waves with incredible accuracy. A blurred echo indicates a rough surface, such as a moth. A sharp echo indicates a smooth object, such as a flying beetle. Some moths can detect bat sonar and will plummet or fly erratically

Evening bats (family Vespertilionidae) over a marsh

when being hit by a bat's sound waves.

The natural curiosity that bats evoke in humans should be resisted; they should not be handled with bare hands, as they are known to carry rabies. The dried blood and urine of bats should be avoided, and the bite of a bat is reason for concern. Even airborne droplets of urine from bats that were loose in houses have infected humans with rabies. This serious viral disease develops slowly in humans. Vaccination can be effective even after a person had been bitten by an infected bat. Late symptoms of rabies include excessive salivation and restriction of the throat, which causes saliva to dribble out of the mouth. Confusion, pain, and other symptoms may also be involved. Bats may carry rabies without displaying any external symptoms, so beware of getting close to these little mammals.

Most bats of the Valley/ Foothill region belong to the family of **evening bats** (Vespertilionidae). Evening bats lack the bizarre facial features of many other bats. Their eyes are small, and there is a projec-tion called the tragus located at the openings to their ears.

The hunting methods of bats change from one species to the next as well as by season, depending on the habits of their prey. In summer months, the zigzag flight of evening bats accompanies the emergence, at dawn and dusk, of flying insects. Some evening bats, such as the **pallid bat** (*Antrozous pallidus*), scoop scorpions, crickets, and other prey from the ground and carry the meal away to be eaten at their roosts, where discarded fragments of their victims accumulate.

Robust populations of the **hoary bat** (*Aeorestes cinereus*) and the **western red bat** (*Lasiurus blossevillii*) winter along the Pacific coast and migrate inland in early spring. Both bats have long body fur and rounded ears. Their slow wing beats, large size, and fur color help identify them. Hoary bats are yellowish-gray or brown with white-tipped hair, resulting in a frosted look.

As in many other bat species, the young of red and hoary bats (often four in a litter) cling to their mother while she hunts. The combined weight of the

A.

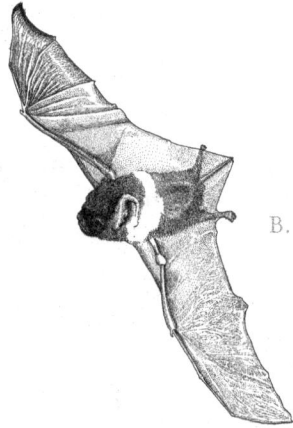

B.

C.

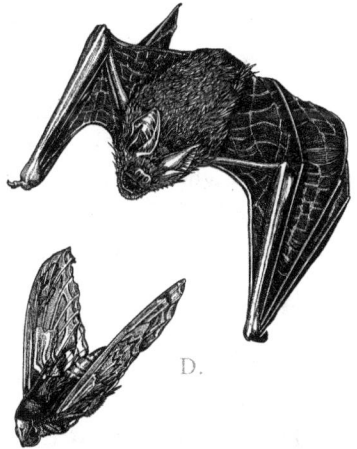

D.

BATS

A. Western red bat (*Lasiurus blossevillii*)
B. Hoary bat (*Aeorestes cinereus*)
C. California myotis bat (*Myotis californicus*)
D. The long ears of pallid bats (*Antrozous pallidus*) reach beyond the muzzle when laid flat.

young can exceed the mother's weight!

Myotis bats (*Myotis* spp., also evening bats) have the distinctively erratic flight and rapid wingbeat often seen in evenings along canals and ponds, and even around street lamps. They roost together under shingles or other such places. Most are some shade of brown. They hunt heavily in summer to store enough body fat to hibernate in winter.

Many species in the free-tailed bat family (Molossidae) are great flyers in terms of both speed and distance, with some flying over fifty miles each night. Their tails extend beyond the tail membrane, and their ears, projecting from the sides of their heads, look strangely like wings. The **Brazilian free-tailed bat** (*Tadarida brasiliensis*) found in the Valley/Foothill region is the same species that nightly emerges by the thousands from Carlsbad Caverns and other caves in the Southwest. The wingbeat of this brown-to-gray bat is relatively slow, and flight looks directional unless the bat is feeding. Free-tailed bats often hunt together and sometimes cooperate in the pursuit of prey. Colonies of free-tailed bats are sometimes found under large wooden bridges.

Rabbits and Hares
(order Lagomorpha)

These leaping mammals have chisel-like incisors that, because they never stop growing, must be worn down by chewing. Behind each of the incisors is a second, peglike tooth.

After rabbits eat, they rest in a safe place and defecate green, relatively undigested pellets, which they then eat and digest in safety. Hares, too, have this ability to maximize the amount of nutrients they get from food.

Unlike rabbits, who hide from predators, hares run to escape danger. Their young are capable of walking almost immediately after birth. They nurse less than a month, then change to a diet of grasses and forbs and are ready to mate the following year.

The incredible fertility of lagomorphs is due to the fact that the female releases eggs when mating occurs, virtu-

Desert cottontail (*Sylvilagus audubonii*)

ally ensuring pregnancy. This high reproductive capacity is important, as they are frequently preyed upon and they have many predators.

The **desert cottontail** (*Sylvilagus audubonii*), common in scrubland, has a white tail and lives in brush and burrows. Prior to giving birth, the female cottontail lines a large underground den with fur from around her belly, making a warm nest and exposing her teats for nursing her young. As with all true rabbits, the young are blind and nearly hairless for the first few days after birth. They soon accompany their mother as she feeds on tender grasses and herbs. Litters of six young are common,

and a mature female can have several litters each year.

Cottontails are numerous in favorable locations; their huddled posture is a common sight along rivers and near barns. An endangered subspecies, the **brush rabbit** (*Sylvilagus bachmani*), occurs in a few riverside forests of interior California.

The **black-tailed jackrabbit** (*Lepus californicus*), inhabits open lands and has long, black-tipped ears, a short black tail, and long, powerful hind legs capable of twenty-foot leaps. The name *jackrabbit* is believed to come from "jackass rabbit," in reference to the large ears. These are useful for radiating excess body heat and,

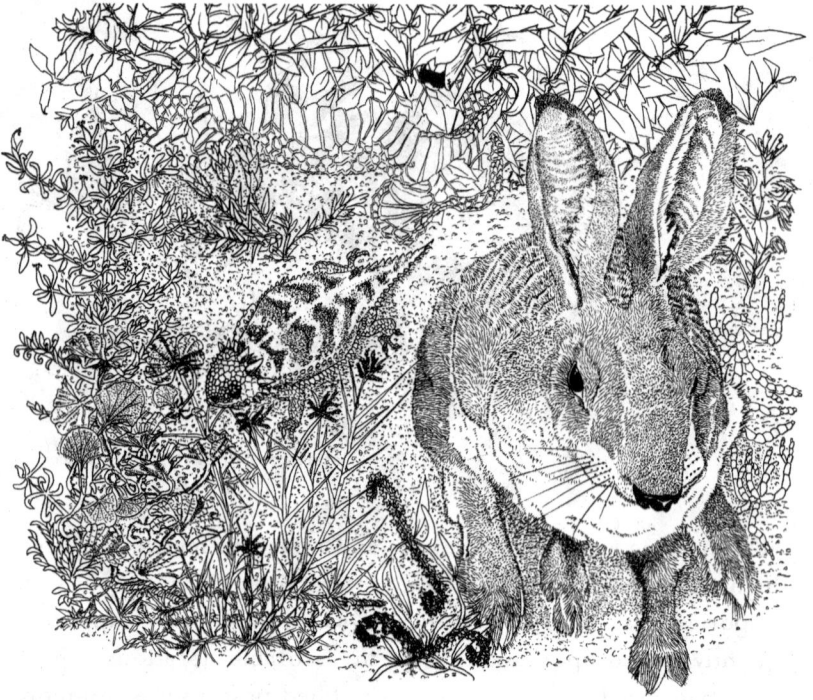

Black-tailed jackrabbit (*Lepus californicus*) hiding under a saltbush

because sound does not travel as well in warm air as it does in cool air, as an adaptation to the warm climates that jackrabbits prefer.

The jackrabbit does not burrow to protect itself—it runs at speeds up to thirty-five miles per hour to avoid coyotes, hawks, wild dogs, and humans. Another strategy hares use is to sit quietly with ears flattened against their backs. These animals frequently rest at the same protected spot near a bush or clump of grass (called a lie), exploding out at a gallop when flushed. Flushing a jackrabbit from a lie is a startling demonstration of just how well their coloration and immobile behavior keep them hidden.

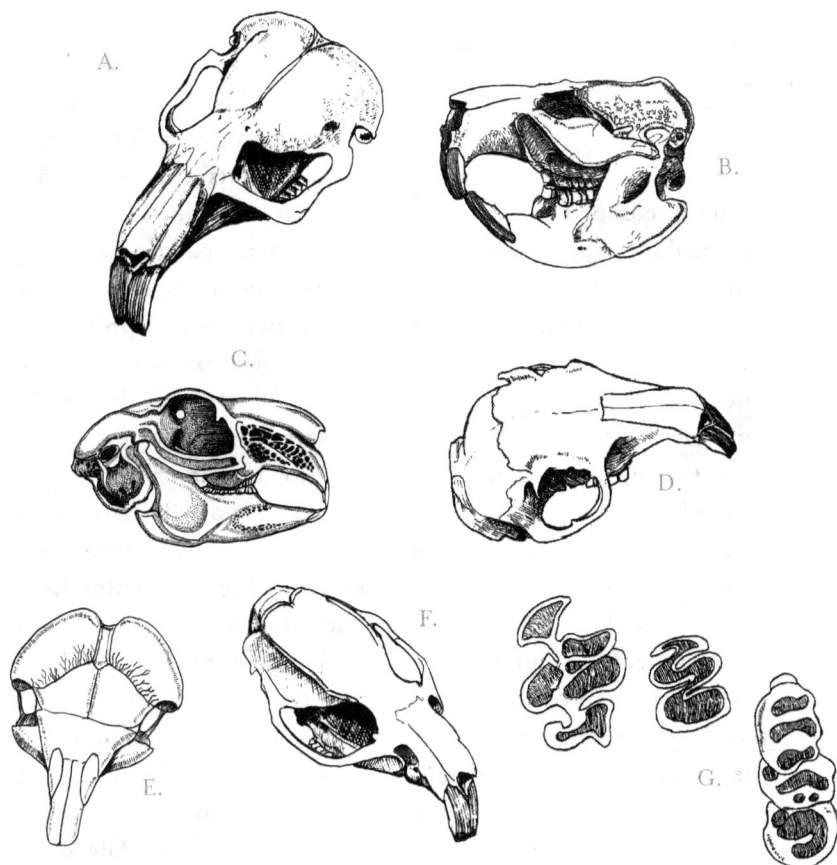

HERBIVORE SKULLS

A. Gopher skulls look flattened, and the zygomatic arch (cheekbones) is wider than the base of the skull. The skull is usually one to two inches (2–5 cm) long.

B. One of the largest rodents in the world, the beaver has a heavy skull that is sometimes over four inches (10 cm) long. Jaw muscles slide along in front.

C. Look closely inside a rabbit's mouth and you will see a second row of peglike incisors lying up against the frontal incisors.

D. On squirrels and marmots, muscles attach to a pointed bone, called the postorbital process, that extends over the eye socket.

E. Massive auditory bullae (ear chambers) help kangaroo rats and pocket mice hear their enemies approaching.

F. On the skulls of native mice and rats, muscles pass through a small triangular opening located in front of the eyes. Their teeth have only two rows of cusps (crowns).

G. With a magnifying glass, it is simple to identify Old World rats by their teeth, which have three cusps and look somewhat like human teeth.

Rodents
(order Rodentia)

Rodents comprise the largest mammalian order, nearly one third of all species of mammals. They range in size from less than mouse-sized to over one hundred pounds (the capybara of South America), with the majority being less than a foot long.

Rodents have continuously growing incisors followed by a large gap before the back, grinding teeth. Their incisors have a resistant enamel layer on the front surface. The rest of the tooth wears faster than this front surface, resulting in the familiar chisel shape we associate with rodent teeth.

The jaw joints of rodents are quite shallow, accommodating a grinding motion for efficient feeding. Their reputation for extraordinary gnawing ability is supported by reports of rats gnawing through twelve inches of concrete.

The **North American beaver** (*Castor canadensis*), found in the Valley/Foothill region and much of the rest of the continent, is the second largest rodent in the world and the largest rodent in North America, weighing from forty to sixty pounds. Its paddle-shaped tail—more a rudder and diving plane than a paddle—is unique among mammals worldwide. It also functions as a prop when beavers feed on land, and the smack of a tail on the water's surface communicates a warning to neighboring beavers. Ears and nostrils have valves that can be closed to exclude water during underwater dives, and the beaver's eyes have transparent, protective eyelids. The five toes of each hind foot are webbed, and the two inside toes have double nails that serve as combs to keep the fur smooth and clean. Clean fur traps more insulating air than ungroomed fur, and a beaver spends much time grooming its fur in order to stay comfortable in cold water. Beavers also use these nail combs to remove splinters of wood from between their teeth.

Beavers eat the inner tissue (cambium) of such streamside trees as willows, alders, and cottonwoods. Like all mammals, they cannot digest wood, and they rely on microbes inhabiting their guts to perform this task. Beavers are famous for constructing dams from sticks, but in this region they often excavate burrows in riverbanks. Beavers mate for life, typically producing one litter per year of two to six young.

Nelson's antelope ground squirrel
(*Ammospermophilus nelsoni*)

Beaver (*Castor canadensis*)

The hot summer temperatures of the San Joaquin Valley present a formidable challenge to diurnal animals. **Nelson's antelope ground squirrels** (*Ammospermophilus nelsoni*) have adapted to this harsh environment by being remarkably tolerant of heat, not even trying to cool off until temperatures are above 107°F. Then they move into burrows and press their bellies and legs against the cool floor. They then resume foraging on the hot surface,

returning to the burrow every fifteen minutes or so. Squirrels lack sweat glands, so when it is extremely hot they wet their heads and chests with saliva, a more efficient form of evaporative cooling than panting.

The **California ground squirrel** (*Spermophilus beecheyi*) has a salt-and-pepper appearance, with a lighter mantle on its shoulders. This small rodent lives in burrows established by extended families.

California ground squirrel
(*Spermophilus beecheyi*)

The numerous dens and exits of their burrows help these squirrels to escape snakes, weasels, and other predators that hunt underground. Ground squirrels will kick sand at snakes to defend themselves and their young. When enemies approach, they emit a chirping whistle. They feed mostly on plant material but include carcasses, insects, and other animal matter in their diets. There were fewer of these native squirrels back when hawks, coyotes, and other natural predators roamed this land. They can be pests, especially when they eat crops or form holes that are a danger to livestock.

The **eastern fox squirrel** (*Sciurus niger*), a large tree squirrel, was introduced to the West. Most are reddish brown with an orange tint on the underbelly. These are the squirrels most frequently seen in city parks and towns. Often they travel from tree to tree, sometimes using overhead wires as tightrope highways. Fox squirrels forage on the ground, often receiving handouts in parks. They are capable of delivering a serious bite. They eat eggs and have affected the population of native songbirds.

Eastern fox squirrel (*Sciurus niger*)

Western gray squirrels (*Sciurus griseus*) are busy during the warm months, when they must stash food away for the winter. These native squirrels bury acorns several inches underground and later locate them with their sensitive noses. They prefer acorns but will consume all sorts of food, including mushrooms, pine nuts, insects, and even bird eggs. When on the ground, this squirrel is vulnerable to its many enemies, which include coyotes, hawks, and bobcats. Two litters per year of two to three young are born after a ninety-day gestation period. They often den in old woodpecker holes or in leaf nests in trees.

Western gray squirrel
(*Sciurus griseus*)

The bubonic plague bacterium—the same species that has killed millions of people worldwide—lives in squirrel fleas. Do not go near dead squirrels, because fleas may hop to your warm body. Also, avoid stray cats that may have eaten a squirrel; cats have infected humans with plague as well.

The lifestyle of the **Botta's pocket gopher** (*Thomomys bottae*) centers around digging. These gophers live almost entirely underground. Pow-erful shoulders and long claws are important digging tools, but their large, ever-growing incisors, typical of rodents, are also used for digging. A flap of skin behind the incisors prevents dirt from entering the mouth and makes it possible for a gopher to show its front teeth with its mouth closed.

Life in a tunnel makes other demands on the gopher. A short tail allows this remarkable animal to back up quickly. Gophers tolerate oxygen levels lower than would be expected for their body size. This is because the opening to a gopher's burrow is usually plugged with soil to keep predators out. A small pelvis allows this remarkable animal to turn in a narrow tunnel by flipping, but this feature creates a different problem for the gopher: the birth canal is too small. The solution is produced by hormones during pregnancy that change the pelvic opening from an "O" to a "U" shape.

Gophers are herbivores, feeding on roots, tubers, and other plant parts. They routinely pull tall plants down into their burrows; the image of the disappearing plant so familiar from cartoons actually does happen. Gophers do not hibernate and must store food for the lean times of the year. Folds of skin on the outside of the pocket gopher's cheeks give it its name and allow it to carry food for storage. Pocket gophers forage outside their tunnels but usually under the cover of darkness and near an escape hole. The most dangerous time in a gopher's life is when it is outside the safety of its burrow, which helps explain their extremely bristly behavior when encountered on the surface.

Despite their common names, kangaroo rats and pocket mice are more closely related to squirrels and gophers than to mice and rats. They have adapted to arid lands by collecting seeds at night, carrying them in external, fur-lined cheek pouches, and spending their days in burrows plugged with soil. They do not require free water because they can derive metabolic water from food, especially fat in the seeds they eat.

Kangaroo rats (*Dipodomys* spp.) are so named because of the way they hop on their large hind legs, counterbalancing with their long, tufted tails. They have large eyes, tawny fur, and a gentle nature, though they can kick with vigor. Their large heads accommodate large chambers in the skull, called bullae, that enhance hearing: a kangaroo rat can actually hear the strike of a predator and take evasive action, an important ability for a small animal that forages after dark in open habitats. They are a favorite item in the diets of many carnivores, especially owls and kit foxes.

Pocket mice are smaller relatives of kangaroo rats. They live a similar but less elusive lifestyle and often forage under shrubs or other vegetative cover. Two that occur in the Valley/Foothill region are the **San Joaquin pocket mouse** (*Perognathus inornatus*) and

Kangaroo rat (*Dipodomys* spp.)

Pocket mice (*Chaetodipus californicus*)

the **California pocket mouse** (*Chaetodipus californicus*).

Features that distinguish kangaroo rats and pocket mice from other rodents are their fur-lined cheek pouches, tawny fur with a light band on each side, fur-tipped tails, and large hind legs.

Most of California's native mice have furry tails and white bellies. They are abundant but rarely seen by humans. The nest of the **western harvest mouse** (*Reithrodontomys megalotis*), woven from shredded plants and lined with cottony seeds of cattail, willow, or cottonwood, is more often seen than the rodent itself. A distinguishing feature of the western harvest mouse is the deep groove in each of its incisors.

The **deer mouse** (*Peromyscus maniculatus*) is an appealing creature with a white belly and feet, large ears, and dark eyes. Feeding on a variety of foods that include seeds, fruits, insects, and plants, deer mice live only a few years, during which time a female may give birth to several dozen young.

If not for predators like hawks, coyotes, owls, and snakes, we would be overrun by rodents.

California voles, or meadow mice (*Microtus californicus*), are among the most

Deer mouse
(*Peromyscus maniculatus*)
California vole, or meadow mouse
(*Microtus californicus*)

A western harvest mouse (*Reithrodontomys megalotis*) and its grassy habitat

prolific of the rodents. They breed year-round and have litters of three to ten. One day after giving birth, a female may ovulate and breed again. Their gestation is just three weeks. The young are weaned in two weeks and can breed within one month. These rodents sometimes occur in such large numbers that they cause traffic accidents when the roads become slick from their crushed bodies.

The dense, brown fur of voles obscures their small ears. Their tails are relatively short and their bodies are stocky. Voles are especially abundant in early spring, when food is plentiful. Active day and night, they search in damp, grassy places—creating tunnels in the grass for safety—for seeds, plant bulbs, and insects.

The **dusky-footed woodrat** (*Neotoma fuscipes*) has a furry tail and white underparts. It constructs a stick nest several feet tall with an assortment of entrances and inner chambers. This nocturnal rodent, like other members of its genus, is a pack rat, always ready to trade twigs or pieces of bark for coins, forks, watches, or other shiny, more interesting items.

The **muskrat** (*Ondatra zibethicus*) is named for a pair of musk glands located near its tail that are used for communication, possibly marking territories and leaving breeding signals. Found in marshes, slow-moving streams, canals, and gravel-pit ponds, muskrats eat bark, marsh plants, and small animals such as clams, frogs, and fish. Their feet are partially webbed, and these rodents are excellent divers, able to remain underwater for longer than fifteen minutes. Their tails, slightly flattened side to side, serve as rudders. Their ears and nostrils close when they dive, and their mouths close behind their incisors so that they can gnaw underwater.

Muskrats dig their burrows in riverbanks or construct shelters like beaver lodges out of cattails and tules. The entrances to their dens are only accessible by diving, a defense against enemies such as bobcats and coyotes that are poor swimmers. Because the name *muskrat* lacks glamour, the valuable fur is sometimes sold as "Hudson seal."

Muskrats (*Ondatra zibethicus*) are about the size of a small, short-legged house cat.

Unlike native mice, foreign rodents are easily identified by their gray bellies and hairless, scaly tails. These common pests were introduced from Europe, probably in storage containers on the ships of early colonists. They infest our houses and garages, building up their populations until a trip to the hardware store for snap traps ends their colonization attempt.

The English word *mouse* comes from the Sanskrit word *musha,* which translates to "thief" and is an accurate summary of the relationship of the **house mouse** (*Mus musculus*) to humans. Unlike most native mice, these rodents sneak around in human dwellings in the dark, eating crumbs or gnawing their way into the greater wealth of stored foods. A fairly consistent but little reported feature of the house mouse is the strong, unpleasant odor associated with it.

Much harm is done by introduced **Norway rats** (*Rattus norvegicus*) and **black rats** (*Rattus rattus*). Much money is spent rat-proofing buildings, and yet they still manage to raid stored grain and other goods and

House mouse (*Mus musculus*)

destroy property by gnawing walls and electrical insulation, sometimes causing fires. They carry several serious diseases, including bubonic plague and trichinosis. Many of them live in towns, nesting in abandoned buildings, burrows, palm trees, and other sheltered places and eating garbage, fruit from backyard trees, and dog food from patios.

Carnivores
(order Carnivora)

Carnivores use their canine teeth for gripping and holding prey. Their robust claws are useful for fighting, digging, climbing, and snagging the skin of prey. Their hunting lifestyle requires that they be alert, strong, swift, and fairly intelligent. Most have keen senses, including a well-developed sense of smell. Though they are distinguished from other mammals by their animal diet, some carnivores eat more plants than meat.

The eyes of carnivores are on the fronts of their heads, causing an overlapping field of vision that enhances depth perception, vital in stalking and striking. By contrast, herbivore eyes are typically on the sides of their heads, increasing the overall field of vision at the expense of depth perception.

Carnivores usually have a single litter of offspring each year. Extended parental care allows the young to acquire the knowledge and skills required to survive by hunting. Many carnivores are solitary but are sometimes mistaken for "packs" when mothers and their young are seen together. Less numerous than their prey, a natural result of the energy transfer along food chains, carnivores are of ecological importance, providing resistance to rapidly increasing populations of rodents.

Raccoons (*Procyon lotor*) feed on fruits, insects, crayfish, eggs, frogs, mice, garbage, and more. They usually weigh about fourteen pounds, with the rare giant in the vicinity of fifty pounds. They use their nimble, handlike front paws to wash food, and the water enhances their sense of touch. Raccoons are probably heard more than they are seen—their vocalizations are variable but usually include a soft, birdlike "chirr." And the distinctive tracks of these nocturnal animals are more often seen than the raccoons themselves.

Raccoons are solitary, except during their breeding season. When one raccoon encounters another, threats and posturing ensue, with the usual result being the retreat of both animals. Trees are the preferred den sites, and mothers raise the young, leading

Raccoon (*Procyon lotor*)

them nightly to stream banks where they learn to use their sensitive fingers to feel for prey. Raccoons are routinely seen begging and marauding around campsites. Resist the urge to feed these masked beggars and secure your food well.

The **ringtail** (*Bassariscus astutus*), a cousin of the raccoon, is also called a ringtail cat or miner's cat, apparently because of its beneficial role in cap-

turing mice around settlers' homes in the 1800s. Ringtails are nocturnal mammals of the foothills with a broad diet that includes small animals and berries.

The **black bear** (*Ursus americanus*) is the only bear species left in California; the California grizzly became extinct early in the twentieth century.

People have considered bears to be almost mystical

Dusky-footed woodrat (left, *Neotoma fuscipes*) and ringtail (*Bassariscus astutus*)

throughout history, perhaps because bears seem similar to humans in so many ways. A bear's track shows a classic plantigrade stance that looks much like our own, although we lack the robust claws. Like ours, the bear's diet varies with the season and may include fish, fruits and other plant parts, and small animals. A skinned bear looks shockingly humanoid. Bears are not shy about chasing other animals away from a desirable piece of food—in fact, they seem to prefer this over actually hunting down their own prey. And bears also crave honey.

Bears do have a more regimented relationship to the calendar than humans, however. They are born in winter, and the arrival of two or three cubs often wakes the mother up from hibernation. While only partially awake, she nurses her young for the next several months. During this time, she may lose up to 40 percent of her body weight. This drastic weight loss makes it crucial that pregnant bears gorge themselves during the fall. During the warm months, bear cubs follow their mothers closely, watching them forage for food. This is a time when mother bears can be ferociously protective; there are nineteenth-century accounts of mother black bears fending off grizzlies.

Named for its bobbed tail, the **bobcat** (*Lynx rufus*) looks like a large, tawny, spotted house cat with a short tail, small ear tufts, and large feet. A ruff of fur on each cheek makes its face looks wide. Most bobcats weigh about twenty pounds, but there are records of males over fifty pounds. Bobcats are extremely wary and have keen senses. Should you ever be fortunate enough to view a bobcat, savor the experience.

Like most cats, bobcats have retractable claws that do not show in their roundish tracks. They hunt by stealth and ambush, often lurking by a game trail, sometimes dropping on their prey from above. Bobcats also stalk prey, relying on sharp eyesight, silent padded feet, and a camouflaged coat to get near enough for a quick spring. They have a reduced number of teeth, specialized for piercing and crushing the necks of their

Bobcat (*Lynx rufus*)

prey. Bobcats will attack animals from the sizes of birds and mice to those of young deer or sheep. They are particular eaters; some refuse to eat anything that isn't freshly killed. During spring mating rituals, partners sound much like yowling house cats. The female raises two to three kittens alone. Den sites are dry and secluded and include cracks in rocks, hollow trees, and burrows. A mother bobcat brings live rodents to her young to teach them to hunt and kill.

Our largest native cats, **mountain lions** (*Puma concolor*), sometimes called cougars, will take nearly anything, from mice to deer. They even prey upon porcupines, skunks, and feral cats. An adult lion eats about one deer a week and will bury what it doesn't consume; a buried, stinking carcass is a sign that one of these big cats is around.

A mountain lion's favorite attack method is to slowly stalk its prey and then make a powerful leap, sinking its canine teeth into its victim's neck or attempting to suffocate it. Mountains lions are not invincible and are sometimes seriously injured when hunting. There are even tales of mountain lions fatally stuck to the ground by the antlers of the elk they had killed.

Although mountain lions

Mountain lion (*Puma concolor*)

a mountain lion, a human in this posture looks like prey.

Coyotes (*Canis latrans*) often call to each other at night with a series of high-pitched yips and howls, a serenade known as the song of the West. The coyote often hunts at dusk, when it ranges over its territory and chases down or carefully stalks small animals, including gophers, mice, rabbits, and insects. Fruits and other plant material are also important in the diet of this versatile animal.

Coyotes are usually over three and a half feet long and weigh twenty to fifty pounds, but there are records of coyotes weighing over seventy-five pounds. They are intelligent and have survived attempts to destroy their species with traps, guns, and poison. Their adaptability is legendary. At one time, there was a bounty for coyotes, but today they are respected for their role in killing rodents. Coyotes often hunt cooperatively in pairs, taking turns pursuing hares that broadly circle when fleeing danger.

Coyotes den in burrows that they sometimes dig or enlarge themselves. Litters typically

hunt deer, the big cats also benefit their hoofed prey, thinning out herds by preying on slow and weak individuals. The cats keep deer moving during winter migrations, which reduces overgrazing while strengthening the herd. There are a few cases of mountain lions hunting down humans. At particular risk are children playing alone, bending over and digging; to

Coyote (*Canis latrans*)

contain five pups that both parents feed and teach. Coyotes sometimes interbreed with dogs.

Kit foxes (*Vulpes macrotis*) are found on the arid lands of the San Joaquin Valley. Because much of their habitat has been lost to development and agriculture, kit foxes are no longer a common sight in evenings when they emerge to forage for rodents, rabbits, insects, lizards, and fruit. Kit foxes have numerous dens throughout their home range. They move in a gliding, ghostlike manner, effectively using whatever cover is available. Humans and coyotes are major enemies of kit foxes.

Gray foxes (*Urocyon cinereoargenteus*) are unusual in that they climb trees, an ability aided by short, curved claws. They favor leaning or low-forked trees, and there are reports of these foxes taking refuge in old nests of hawks and crows.

The gray fox has a salt-and-peppery gray back, orange on its legs, and white underparts. The hind feet of foxes land in the imprints of their front feet, like those of cats. This trait increases their stealthiness, reducing the possibility of stepping on noisy ground cover. Gray foxes are active hunters, locating and stalking their prey with sharp vision, acute hearing, and sensitive noses. Their diet is mostly rodents, rabbits, and birds, with insects, carrion, eggs, and a variety of fruits making up the balance. Gray foxes cache food, using their paws to dig holes and their noses or muzzles to cover them.

Foxes use large musk glands near the base of their tails to mark territory. They often leave scat in conspicuous locations, like in the center of a trail or on rocks.

Three to five young foxes are born in a den in the spring. Both parents provide food, but only the protective mother shares the den with the young. The mother's ability to protect her young is enhanced by her ability to spot a motionless human.

Gray fox (*Urocyon cinereoargenteus*)

An awesome combination of alertness, appetite, stealth, and energy, the **long-tailed weasel** (*Mustela frenata*) is a perpetually hunting streak of chestnut brown. Flowing through its environment, curiously investigating every promising odor, the weasel is the terror of the rodent world. Its sleek body, short legs, and small head allow it to easily invade rodent burrows, where it kills prey with a bite to the base of the skull.

Weasels eat rodents, rabbits, eggs, birds, and other small creatures. They sometimes raid poultry houses, but they prefer rodents. Contrary to popular opinion, it is not bloodthirstiness that motivates them to kill so prodigiously. Their high metabolic rate makes them high-strung and constantly hungry: they eat about 40 percent of their body weight daily.

Male weasels can be over twenty inches long, and females are about one foot in length. This difference is size allows the sexes to reduce competition by hunting different-sized prey. Weasels den in old rodent burrows, sometimes lined with the fur of the former owner.

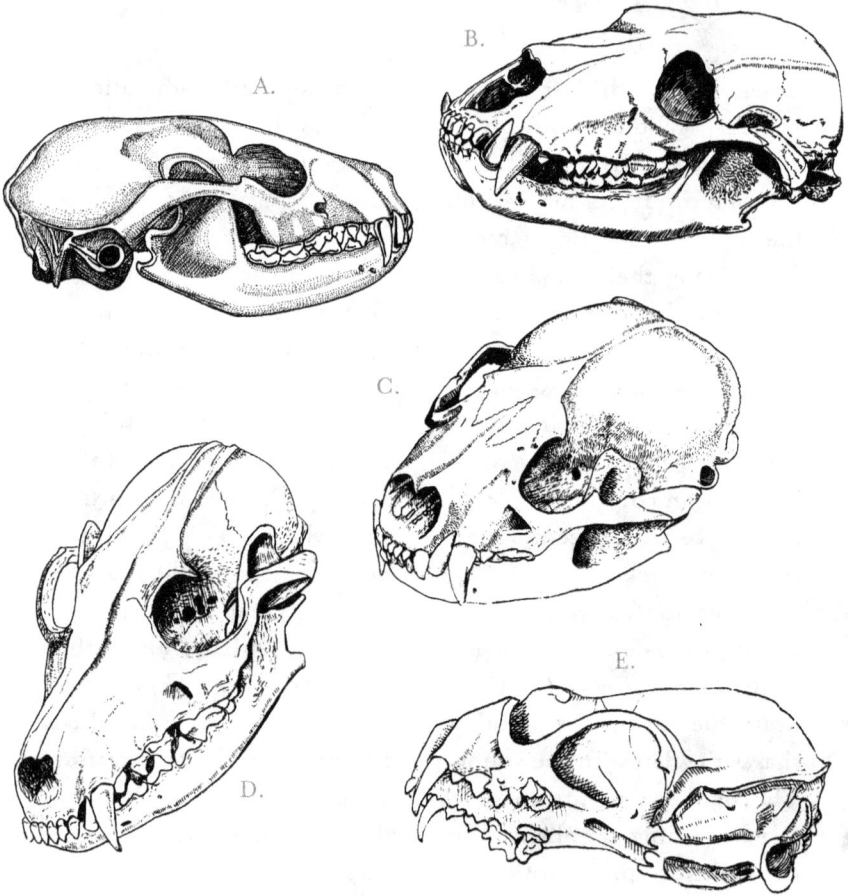

A.

B.

C.

E.

D.

SKULLS OF CARNIVORES

A. Raccoon skulls look like dogs' skulls, but the snout is shorter. Usually there are 20 teeth in the upper jaw. Raccoons have slicing premolars and grinding molars

B. The robust skulls of bears are over eight inches long (20 cm). Canines and molars are thick. There is a brow-ridge over the eye sockets.

C. Mountain lions, bobcats, and their relations have large canine teeth, small incisors, and only 6 to 8 upper-cheek teeth. Cats are strict meat eaters with no need for flattened molars to grind plant materials.

D. Dogs, foxes, coyotes, and their relations have 18 to 20 teeth in the upper jaw. The molars have both flat and pointed grinding surfaces.

E. Weasels, badgers, and their relatives usually have 18 or fewer teeth in the upper jaw. The upper rear molar looks squashed or dumbbell-shaped. The braincase is long and the snout is short.

Long-tailed weasel (*Mustela frenata*)

ferocity is far more important for defense. For example, there is the story of a hawk that captured a weasel and was killed by its prey in mid-air; the hawk was later eaten by the weasel that was to have been its meal.

The **American badger** (*Taxidea taxus*) hunts mostly at night, waddling about and sniffing burrows; its sense of smell is keen enough to distinguish between a vacant burrow and one that is occupied. Having, as Ernest Thompson Seton said, "bartered speed for strength," the badger cannot climb trees well or run fast, but it can dig at a fantastic rate and corner its rodent prey in their own underground burrows. We know a group of eleven people who once attempted to dig out a digging badger but ended up with only four hours of sweaty exercise and a hole twenty feet long and six feet deep.

Powerful shoulders and long claws on the forelimbs serve as the badger's shovel. This member of the weasel family has a heavy, short-legged body about thirty inches long with a muscular neck and a white line on the face. Thick hide and long fur around the face and

There are four to nine young in a litter.

Like skunks, weasels have musk glands near the base of the tail, but they do not spray musk. Although their small size is helpful for hunting, the advantage is balanced by increased vulnerability to predators. Their black-tipped tails may serve to divert the attacks of enemies, but their

American badger (*Taxidea taxus*)

shoulders ensure that minimal damage occurs to the badger in its brief struggles with its prey. Badgers' burrows are elliptical, accommodating their flat bodies. Mothers bear litters of two and raise them alone.

Because badgers are unwary, depending on their formidable strength to defend themselves, they are often trapped, shot, poisoned, or run over. This is unfortunate, because their rodent diet is a benefit to humans. Badgers are now very rare in the Valley and are seen more often in the surrounding hills.

The striped skunk (*Mephitis mephitis*), another member of the weasel family, is not very fast or very cautious, perhaps because its weapon, so like tear gas, offers an effective defense. Skunks, especially juveniles, are often hit on the highway, perhaps because they do not believe that anything would attack them. Because of this unwary attitude, striped skunks are often encountered by people who are out walking at night; a heavy human footfall would cause most wild mammals to flee, but the skunk, confident of its safety, will often continue foraging. Among wild animals, skunks are the number-one carrier of rabies in the United States.

Skunks search at night for insects, mice, eggs, and fruit and usually spend their daylight hours in burrows or hol-

Striped skunk (*Mephitis mephitis*)

Typically, five or six young are born, but litters of ten are known. The mother raises the young alone, leading nightly foraging parties in an amusing black-and-white parade.

The range and habitat of the **western spotted skunk** (*Spilogale gracilis*) overlap with those of the striped skunk. A teacup-shaped spot on its forehead and a distinctive pattern, more striped than spotted, identify this kitten-sized mammal. It dens under the floors of old buildings, in abandoned burrows, and among rock piles. The female bears three to seven young in the spring and raises them alone.

low logs. They often line their burrows with grass.

If confronted by an enemy, a striped skunk begins its warning ritual: foot stomping, tail raising, false charges, and backing toward the enemy. If this is not effective, it sprays musk toward the aggressor from glands near the anus. The spray can be accurate to fifteen feet and cause extreme irritation of the enemy's nose and eyes, resulting in nausea and temporary blindness. Some sources report that the mist from the spray can carry up to thirty feet.

One of the warning postures of the spotted skunk (*Spilogale gracilis*)

Like all skunks, the spotted skunk is boldly marked in black and white as a warning of the potent, malodorous weapon it bears. Further warnings are given when this skunk is threatened: first it stamps its forefeet petulantly, then it raises its tail, prominently displaying the white tip like a pom-pom, and finally it does a handstand with its tail arched over its head. From this acrobatic posture, the spotted skunk will spray if further pressed. Its musk is reported to be more intense but of shorter duration than that of the striped skunk, although both are foul enough to make the comparison moot.

Generally the spotted skunk is quite calm and may become accustomed to the presence of people nearby while it forages in barns for mice and insects. More agile than larger skunk species, spotted skunks climb to forage or as an escape. Very inquisitive, especially where food is concerned, they make nightly rounds to forage for the rodents, insects, reptiles, eggs, fruit, and vegetables that make up the majority of their diet. This skunk eats more rodents than its striped cousin.

Humans, foolish dogs, and great horned owls are the spotted skunk's major enemies.

Otters, yet another member of the weasel family, have had bad luck with humans in the past. Mollusc hunters, believing that sea otters (*Enhydra lutris*) ate more than their fair share of abalone, often shot them on sight. Meanwhile, fishermen were shooting **North American river otters** (*Lontra canadensis*), alleging that they ate too much trout and salmon. Both species were trapped and hunted for their thick fur—otters have nearly one thousand hairs per square inch of skin! In danger of going extinct, these playful and curious carnivores attracted public attention and were given protected status. River otters hunt for clams, crayfish, and fish along a few inland rivers in the Valley/Foothill region.

North American river otter (*Lontra canadensis*)

Hoofed Mammals
(order Artiodactyla)

Be it pig, goat, sheep, cow, or antelope, the body of a hoofed mammal is designed for mobility. Much of the bone mass is positioned near the trunk, for muscular power. Feet are reduced to just a few bones and toes, and the hooves that envelope the toes are lightweight, composed of the same keratin protein found in human fingernails. The heel bone, or calcaneus, is positioned up where the knee would be expected, which also reduces weight at toe tips.

Most hoofed mammals have four toes, with two on the ground and two dewclaws that touch the ground only in soft mud.

Unlike all the other hoofed mammals, **wild boar** (*Sus scrofa*), or wild pigs, have enlarged canine teeth, or tusks, used for digging and defense. Wild pigs in California are the descendants of domestic pigs that escaped or were released. Their habit of rooting for food damages wild habitats, but hunters have kept their population somewhat controlled.

Horns are permanent features found on cows, goats, sheep, and other members of the family Bovidae. Antlers, on the other hand, are shed annually. These branching, bony structures are found in the family Cervidae (deer,

Wild boar (*Sus scrofa*)

elk, moose). Most grow in spring, covered with velvet, then mature in autumn and fall off in winter. Antlers are important in the fall rut, in which males spar for access to females. It is not true that the number of points on a buck's antlers corresponds to his age.

The vast herds of **tule elk** (*Cervus elaphus nannodes*) that once roamed California grasslands have been, during the process of settlement by Euro-Americans, almost completely eliminated. Remnant herds can be seen grazing inside various reserves, such as Tupman in the south and Los Banos further to the north, and efforts to preserve this species from extinction led to the establishment of a free-ranging herd on the east side of the Sierra Nevada, in the Owens Valley.

Tule elk are gregarious, with cows and calves forming herds separate from male herds in the summer. In the fall rut, bulls attempt to take over cow herds as harems. As other bulls challenge those with harems, high-pitched bugling rings through the air. Combat is rare; posturing dominates the competi-

Tule elk (*Cervus elaphus nannodes*)

tion, and the larger, fitter bull generally wins. Elk are grazers, eating grasses and similar herbaceous plants, though they do sometimes browse on the leaves of shrubs and trees.

Mule deer (*Odocoileus hemionus*), so named because their long ears (sometimes over ten inches long) resemble those of mules, are the most important big game animals in California. Their predators also include cougars, bears, coyotes, and bobcats. Other carnivores sometimes kill deer, especially fawns and weak individuals. A

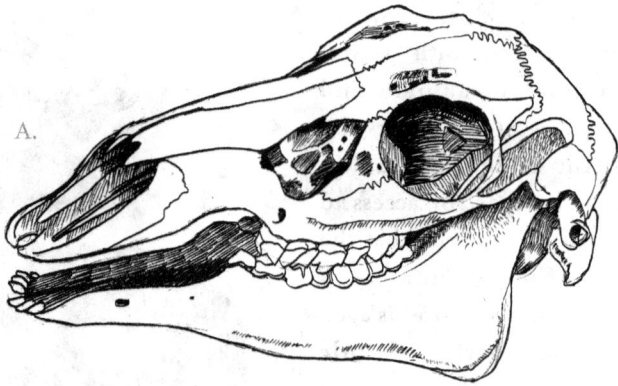

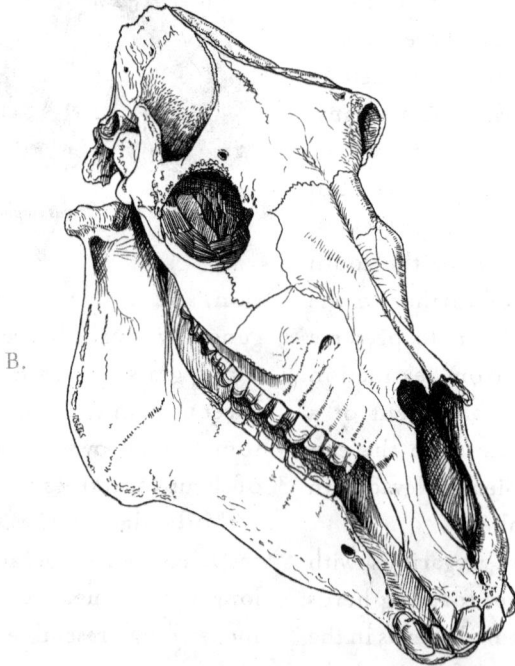

SKULLS OF HOOFED MAMMALS

A. Deer, sheep, goats, and their relations have a toothless pad on the upper palate and teeth on the lower jaw that occlude. This allows them to tug on plants rather than to clip them.

B. Donkeys, horses, and their relations have a massive jawbone, and the skull is often longer than a human forearm. There are large teeth on both jaws.

Mule deer (*Odocoileus hemionus*)

single cougar can kill up to fifty deer annually.

Mule deer are predominantly browsers, eating leaves from shrubs and trees, although grazing is seasonally important to their diet. They are not true herding animals, although they often group where food is abundant.

Pronghorns (*Antilocapra americana*) are commonly called antelopes, but they are actually in a separate family, Antilocapridae. Their horns are unique in that they are branched, with the resulting prong being the namesake of this animal. They are the fastest North American mammals, able to sustain speeds of fifty miles per hour

and reach sixty miles per hour for short bursts. Pronghorns have been known to race cars, seeming to flaunt their great speed.

Geographical features in the Valley/Foothill region named for these fleet animals, such as Antelope Valley, attest to their past abundance. Pronghorn once formed large herds here, but they were extirpated by hunting and habitat loss. The last report of pronghorns in western Fresno County was made in 1954.

Pronghorn (*Antilocapra americana*)

INDEX

A

Acorn, 8–10, 184
Acorn woodpecker, 8, 181, 184
Admiral, Lorquin's, 109
Africanized honey bee, 114
Agaric, yellow staining, 60
Alder, white, 19
Alfalfa, 111
Alfalfa leafcutter bee, 115
Alkali mallow, 24
Alkali sacaton, 36
Alligator lizard, southern, 151
Amanitin toxin, 54
Amaranth, 30–31
American avocet, 171–172
American bittern, 165
American cockroach, 91
American crow, 191
American house spider, 78
American kestrel, 177
American shad, 122
American wigeon, 169
Amphibian anatomy, 134
Angelwing, 108
Anise swallowtail, 108
Anna's hummingbird, 182, 187, 188
Ants, 115

Ant, velvet, 112
Ant lion, 98
Antelope, 245
Antlers, 242–243
Aphids, 97
Aquatic plants, 18–23
Arachnids, 74
Arrowhead, 20–21
Artist's conk, 61
Asian clam, 69
Assassin bug, 94
Avocet, American, 171–172

B

Baby blue eyes, 46
Backswimmer, 96
Bacteria, 51, 78
Badger, 236–238
Bald eagle, 174
Barn owl, common, 178–179
Barn swallow, 190
Bass, 128–129
Bats, 210–214
Beak types, 160
Bears, 229, 231, 236
Beaver, 217–219
Bees, 85, 113–115
Beetles, 85, 99–102

Wooly mullein, 28–29
Wren, Bewick's, 182

Y

Yellow jacket, 112
Yellow staining agaric, 60
Yellow star thistle, 29
Yellow-billed magpie, 191
Yellow-legged frog, 142–143
Yerba santa, 15

ABOUT THE AUTHORS

Derek Madden is a professor of biology at Modesto Junior College. He is best known for his discovery of a new species of parasite, and for his work on African ants and giraffes. Madden spent nearly ten years roaming roadsides and habitats of Central California, often illustrating wildlife as cars rushed by, under intense summer sun, or in the rain and fog, to document the life of California's "Outback" region.

Ken Charters completed a degree in biology from CSU Fresno, and he continued his ecological research as a grad student at Northern Arizona University. Charters is currently a professor of biology at Cochise College, where he teaches and studies ecosystem dynamics.

Erinn Madden graduated with a degree in biology from University of California, Davis, and has spent nearly a decade studying local ecology and animal movements through tropical ecosystems. He is currently a biological technician in Davis, California.

www.ingramcontent.com/pod-product-compliance
Lightning Source LLC
Chambersburg PA
CBHW071735270326
41928CB00013B/2683